D1085216

The History of Science

THE HISTORY OF PHYSICS

The History of Science

THE HISTORY OF PHYSICS

Anne Rooney

New York

For my father, Ron Rooney, who introduced me to the wonders of science. Happy 80th birthday—and thank you.

Acknowledgments

With special thanks to Dr Adrian Cuthbert for his expertise in physics, and to Mary Hoffman, Shah Hussain, Sue Frew, and Jacqui McCary for less specific but no less essential support. Thanks, also, to my tolerant and patient editor Nigel Matheson.

Eppur si muove

This edition published in 2013 by:

The Rosen Publishing Group, Inc.
29 East 21st Street, New York, NY 10010

Library of Congress Cataloging-in-Publication Data

Rooney, Anne.
The history of physics / Anne Rooney. -- 1st ed.
p. cm. -- (The history of science)
Includes bibliographical references and index.
ISBN 978-1-4488-7229-9 (library binding : alk. paper)
1. Physics--History. I. Title.
QC7.R68 2013
530.09--dc23

2012009970

Manufactured in China

SL002278US

CPSIA Compliance Information: Batch #S12YA: For further information, contact Rosen Publishing, New York, New York, at 1-800-237-9932

Contents

Introduction: The Book of the Universe 6

Chapter 1 Mind Over Matter 14
The First Physicist? • Atomic and Elemental Matter • The Birth of Solid-State Physics
Atoms and Elements

Chapter 2 Making Light Work—Optics 34
A First Look at Light • Out of the Dark • Wave-Fronts and Quanta
A New Dawn—Electromagnetic Radiation • At the Speed of Light

Chapter 3 Mass in Motion—Mechanics 62
Mechanics in Action • The Problem of Dynamics • The True Birth of Classical Mechanics
Air and Water • Putting Mechanics to Work

Chapter 4 Energy Fields and Forces 82
The Conservation of Energy • Thermodynamics • Heat and Light • Discovering Electricity
Electromagnetism—The Marriage of Electricity and Magnetism • More Waves

Chapter 5 Into the Atom 110
Dissecting the Atom • Quantum Solace • Things Fall Apart
The End of the Classical Atom

Chapter 6 Reaching for the Stars 140
Stars and Stones • From Watching to Thinking • The Earth Moves—Again
The Invisible Becomes Visible • Galileo, Master of the Universe • Cataloging the Skies
Far, Far Away • The Secret Life of Stars

Chapter 7 Space-Time Continuing 182
A Brief History of Time • Everything is Relative • Back to the Beginning
From Cosmic Egg to Big Bang

Chapter 8 Physics for the Future 196
Ripping It Up and Starting Again • Where Next for Matter?

Glossary 204

For Further Reading 205

Index 206

THE BOOK OF THE UNIVERSE

"The book of the Universe cannot be understood unless one first learns to comprehend the alphabet in which it is composed. It is written in the language of mathematics, and its characters are triangles, circles, and other geometric figures, without which it is humanly impossible to understand a single word of it; without these, one wanders about in a dark labyrinth."

Galileo, **The Assayer,** *1623*

Physics is the fundamental science that underpins all others, the tool with which we explore reality; it aims to explain how the universe works, from galaxies to subatomic particles. Many of our discoveries about the physical world represent the pinnacle of human achievement. *The Story of Physics* traces the path of humankind's attempts to read the book of the universe, learning and using the language of mathematics that the Renaissance scientist Galileo Galilei (1564–1642) describes. It also reveals how little we still know—all the physics we have deals with only 4 percent of the universe, the other 96 percent is a mystery still to be unraveled.

The Birth of Physics

Before the development of the experimental method, early scientists—or "natural philosophers," as they were known—applied reason to what they saw around them and came up with theories to explain it. As the heavenly bodies seem to move across the sky, for example, many of our forebears

The Andromeda galaxy is the nearest galaxy to our Milky Way: physics tries to explain everything, from the beginning of time to the end of the universe.

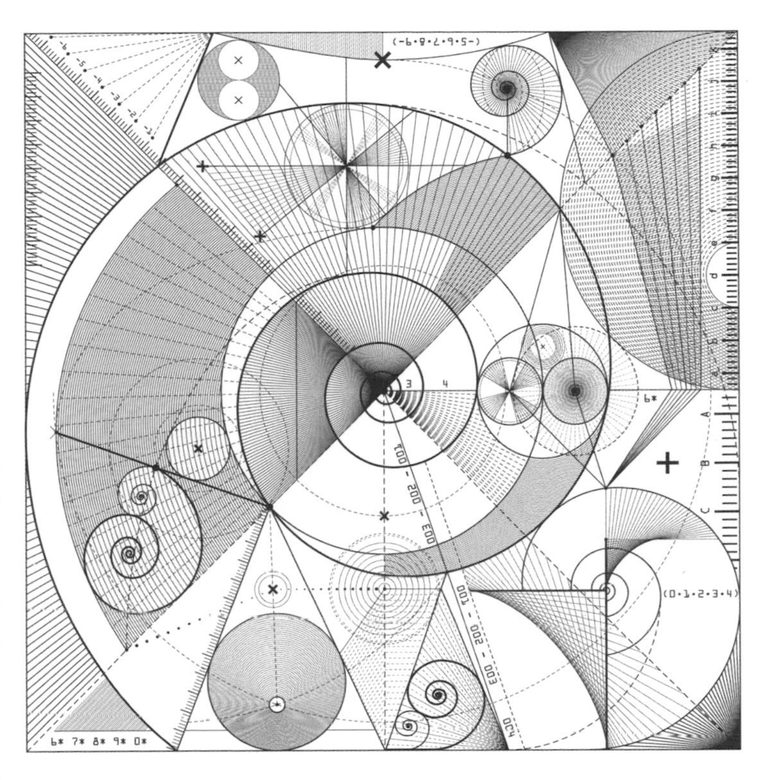

The patterns, shapes, and numbers that structure the natural world are the subject matter of physics.

concluded that the Earth is at the center of the universe and everything revolves around it.

The few who thought differently had to come up with good arguments that would refute the commonsense solution, and for 2,000 years they were outnumbered and sometimes ridiculed or even persecuted.

Many superstitious and religious beliefs have their roots in explaining the observed world. The sun rises because it is driven across the sky by a supernatural charioteer, for example. Science, on the other hand, endeavors to find the real nature and causes of observed phenomena. The Ancient Greeks are the earliest people we know of who tried to replace mystical and superstitious explanations with those based on observation and reason. The first person

THALES OF MILETUS (*ca.*624 BCE–*ca.*546 BCE)

The first named person we can call a scientist and philosopher lived more than 2,500 years ago in what is now Turkey. Thales studied in Egypt, and is credited with bringing mathematics and astronomy to Greece. Considered one of the Seven Sages of ancient Greece, he was reputed to be extraordinarily clever and may have taught the philosophers Pythagoras and Anaximander. Thales suggested that there is a physical rather than a supernatural cause for all phenomena in the world around us, and so began the search for the physical causes that determine how things behave. As none of his writings survive it is difficult to assess his true contribution.

to try to explain the natural world without recourse to religious belief may have been Thales, but the first true scientist was perhaps the Greek thinker Aristotle (384–322 BCE), a thorough empiricist. He believed that by careful observation and measurement we can gain an understanding of the laws governing all things. Aristotle was a pupil of Plato (*ca*.428–347 BCE), who followed a deductive path (see box), believing that reason alone was enough to enable humankind to unravel the mysteries of the universe. Aristotle put his faith in "inductive reasoning"—that is, logic working from observation of the world. He had the beginnings of a scientific method.

Although he did not propose experiments, Aristotle advocated a full investigation of everything previously written on a topic (a literature review, in modern terms), experimental observation and measurement, then the application of reason to reach a conclusion.

The Greeks were the first to divide science into different disciplines. The great library at Alexandria produced the first library catalog, which was essential to the type of literature review that Aristotle proposed should be part of any investigation.

INDUCTIVE AND DEDUCTIVE REASONING

Deductive reasoning is a "top-down" method exemplified in Plato's approach. The scientist or philosopher constructs a theory, evolves a hypothesis to test it, makes observations, and arrives at confirmation (or refutation) of the hypothesis. Inductive reasoning begins with observation of the world and works toward an explanation by identifying a pattern, then proposing a hypothesis to explain it and moving to a general theory. Aristotle's methods were inductive. Isaac Newton (1642–1727) was one of the first to recognize that both deductive and inductive reasoning have a place in scientific thought.

A medieval depiction of Thales

From Empiricism to Experiment

With the end of the Hellenistic age (the height of Classical Greek civilization), the use of the scientific method to understand the natural world declined until the rise of Arab science from the seventh century CE. The brilliant Ibn al-Hassan Ibn al-Haytham (965–1039) developed a procedure

> *"I would rather find the true cause of one fact than become King of the Persians."*
> Democritus (*ca.*450–*ca.*370 BCE), philosopher

similar to the modern experimental method. He began with a statement of a problem, then tested his hypothesis through experiments, interpreted the data, and came to a conclusion. He adopted a skeptical and questioning attitude, and recognized the need for a rigorously controlled system of measurement and investigation. Other Arab scientists added to this. Abu Rayhan al-Biruni (973–1048) was aware that errors and bias could be introduced by faulty instruments or fallible observers. He recommended that experiments should be repeated several times and the results combined to give a reliable result. The physician Al-Rahwi (851–934) introduced the concept of peer review, suggesting that medical men should document their procedures and make them available to other physicians of equal standing—although his principal motivation was to avoid punishment for malpractice. Geber (Abu Jabir, 721–815) was the first to introduce controlled experiments in his field of chemistry, and Avicenna (Ibn Sina, *ca.*980–1037) declared that induction and experimentation should form the foundations of deduction. The Arab scientists valued consensus and tended to weed out fringe ideas that were not supported by others.

Developments in Islam eventually hampered the pursuits of Arab scientists, though. To question the world came to be seen as a blasphemous activity, as if prying into the ways of God and attempting to violate sacred mysteries. The activities that a faithful (or prudent) Muslim scientist could undertake were circumscribed. The torch of scientific endeavor that was cast away by Islam's natural philosophers was then picked up by the medieval scholars of Christian Europe.

Arab science and the works of Aristotle flowed into Europe in Latin translation in the early Middle Ages. The writers of the 12th-century Renaissance period began to integrate the embryonic scientific method into their own studies and to nurture it, but at first did not challenge the Classical authorities. The English Franciscan friar Roger Bacon (*ca.*1210–*ca.*1292) was one of the first to doubt unquestioning acceptance

THE SCIENTIFIC METHOD

The scientific method as generally employed today follows these stages:

- Statement of a problem or question. This may then be narrowed down to something that can be tackled with an experiment or set of experiments.
- Statement of a hypothesis.
- Designing an experiment to test the hypothesis. The experiment must be a fair test, with controlled variables (which remain the same) and an independent variable (the condition that will be varied).
- Carrying out the experiment, making and recording observations and measurements.
- Analyzing the data.
- Stating conclusions and subjecting them to peer review.

of the writings of the ancients, and advocated renewed examination of established ideas. He particularly targeted Aristotle, whose ideas in many areas were accepted as gospel truth, recommending instead that Aristotle's conclusions should be tested. Aristotle would no doubt have approved of the application of empirical methods to re-evaluate and question his writings. In his own scientific investigations, Bacon followed a pattern of constructing a hypothesis on the basis of observations, then carrying out an experiment to test the hypothesis. He repeated his experiments to be sure of his results, and documented his methods meticulously so that they could be scrutinized by other scientists. He called experimentation the "vexation of nature." He said, "We learn more through artful vexation of nature than we do through patient observation."

Another Bacon, the English lawyer and philosopher Francis Bacon (1561–1626), proposed a new approach to science, which he published in 1621 in *Novum Organum Scientiarum (The New Organon of the Sciences)*. He believed that the results of experiments could help to untangle conflicting theories and help humankind move toward truth. He promoted inductive reasoning as the basis of scientific thought. Bacon set out a process of observation, experimentation, analysis, and inductive reasoning that is often considered the start of the modern scientific method. His method begins with a negative aspect—ridding the mind of "idols" or received notions—and progresses to a positive aspect that involves exploration, experimentation, and induction.

The Scientific Revolution

Although Bacon was the first to formulate the method, a similar approach to experimentation had already been adopted by Galileo. Galileo was a great proponent of inductive reasoning, realizing that empirical evidence from a complex world would never match the purity of theory. It is not possible to take account of every variable in an experiment, he reasoned. For example, he believed that his experiments with gravity could never remove the effects of air resistance or friction. However, standardizing methods and measurements means that an experiment carried out

> Bacon is said to have died after conducting an experiment in producing the first frozen chicken in 1626:
> *"As [Sir Francis Bacon] was taking the air in a coach with Dr Witherborne (a physician) towards Highgate, snow lay on the ground, and it came into my lord's thoughts, why flesh might not be preserved in snow, as in salt. They were resolved they would try the experiment at once. They alighted out of the coach, and went into a poor woman's house at the bottom of Highgate Hill, and bought a hen, and made the woman gut it, and then stuffed the body with snow, and my lord did help to do it himself. The snow so chilled him, that he immediately fell so extremely ill ... [he contracted] such a cold that in two or three days, as I remember Mr Hobbes told me, he died of suffocation."*
>
> John Aubrey, *Brief Lives*

repeatedly, perhaps by different people, can produce a set of results from which general conclusions may be extrapolated. Galileo put enough faith in the experimental method to risk his reputation on a public demonstration to settle an argument in 1611. He and a rival professor at Pisa had argued about whether the shape of objects of the same material (and so the same density) affected their ability to float in water. Galileo challenged the professor to a public demonstration, saying he would stand by the results of the experiment; the other professor did not turn up.

Scientific Societies

The growing interest in science gave birth to the scientific societies that sprang up around Europe from the 17th century. These provided a focus for scientific talk, experimentation, and development. The first was the Lyncean Academy (Accademia dei Lincei) formed by Federico Cesi, a wealthy Florentine with a keen interest in science. Although only 18, Cesi believed scientists should study nature directly rather than rely on Aristotelian philosophy as a guide. The first members of the academy lived communally in Cesi's house, where he provided them with books and a fully equipped laboratory. Members included the Dutch physician Johannes Eck (1579–1630), the Italian scholar Giambattista della Porta (*ca.*1535–1615), and—most famously—Galileo. At its height, the academy had 32 members spread around Europe. The academy stated its aims in 1605 as "to acquire knowledge of things and wisdom... and peacefully to display them to men... without any harm." Despite this, the group was accused of black magic, opposing Church doctrine, and living scandalously.

The Lyncean was a very personal venture, and when Cesi died in 1630 it soon folded. It was succeeded by the Academy of Experiment in Florence, founded in 1657 by two former

Robert Boyle as a young man

No contemporary portrait of Robert Hooke survives. A portrait existed in the Royal Society in 1710, but it has been suggested that Newton had it destroyed.

Robert Hooke's Micrographia *revealed the tiny details of life for the first time.*

MICROGRAPHIA:
OR SOME
Physiological Descriptions
OF
MINUTE BODIES
MADE BY
MAGNIFYING GLASSES
WITH
OBSERVATIONS and INQUIRIES thereupon.

By *R. HOOKE*, Fellow of the ROYAL SOCIETY

LONDON, Printed by *Jo. Martyn*, and *Ja. Allestry*, Printers to the ROYAL SOCIETY, and are to be sold at their Shop at the *Bell* in *S. Paul's* Church-yard. M DC LX V.

pupils of Galileo, Evangelista Torricelli (1608–47), and Vincenzo Viviani (1622–1703). It, too, was short-lived, closing after 10 years in 1667 at about the time the center of scientific development moved from Italy to Britain, France, Germany, Belgium, and the Netherlands.

The greatest of the scientific societies was the Royal Society of London. Though officially founded in 1660, its origins lie in an "invisible college" of scientists who began to meet for discussion in the 1640s. At its foundation there were 12 members, among them the English architect Sir Christopher Wren (1632–1723) and the Irish chemist Robert Boyle (1627–91). Wren's opening address spoke of founding "a Colledge for the Promoting of Physico-Mathematicall Experimentall Learning." The society planned to meet weekly to witness experiments and discuss scientific topics, with Robert Hooke (1635–1703) as first curator of experiments. At first apparently nameless, the name The Royal Society first appears in print in 1661, and in the Second Royal Charter of 1663 the Society is referred to as "The Royal Society of London for Improving Natural Knowledge." It was the first "royal society" of any type. It began to acquire a library in 1661 and then a museum of scientific specimens, and still has Hooke's microscope slides. After 1662, the society was granted a charter to publish books, and one of its first two titles was Hooke's *Micrographia*. In 1665 the Royal Society brought out the first issue of *Philosophical Transactions*, now the oldest scientific journal still in continuous publication.

> *"Galileo, perhaps more than any other single person, was responsible for the birth of modern science."*
>
> Stephen Hawking, British cosmologist, 2009

The Royal Society was quickly followed by the Académie des Sciences in Paris in 1666. Members of the Académie did not need to be scientists and at one point Napoleon Bonaparte was president. Great scientific undertakings rapidly became a source of national pride and international rivalry, particularly to the French Republic and Napoleon's France.

Foucault's pendulum in the Panthéon, Paris, provided a dramatic demonstration that the Earth turns on its axis.

The Best Scientific Tool—The Brain

Without recourse to equipment and without carrying out experiments, Aristotle came up with models for the nature of matter and the behavior of bodies under different conditions that worked with what was already known. In the early 20th century, the physicist Albert Einstein (1879–1955) revolutionized physics and the scientific view of the universe using pen and paper alone. Just like Aristotle, he was working from observations of the universe to develop theories, dealing with phenomena that could not at the time be truly investigated by experimentation or even measurement.

Unlike Aristotle, though, and following a practice started by Newton in 1687, Einstein made rigorous use of mathematics to support his arguments and show that his system worked with what was already known. He made predictions that have since been borne out by observation and experimentation. Considerable mathematics is usually applied to testing a new model in physics now, and in that respect modern physicists have the advantage over previous generations. They now have computers that enable them to carry out speedy calculations that would have taken whole lifetimes in the not-so-distant past.

But behind all developments in science, it is the ingenuity and curiosity of human beings that drives progress, as much in today's universities and research labs as in the outdoor academies of Ancient Greece.

Robert Hooke's microscope

CHAPTER 1

Mind Over MATTER

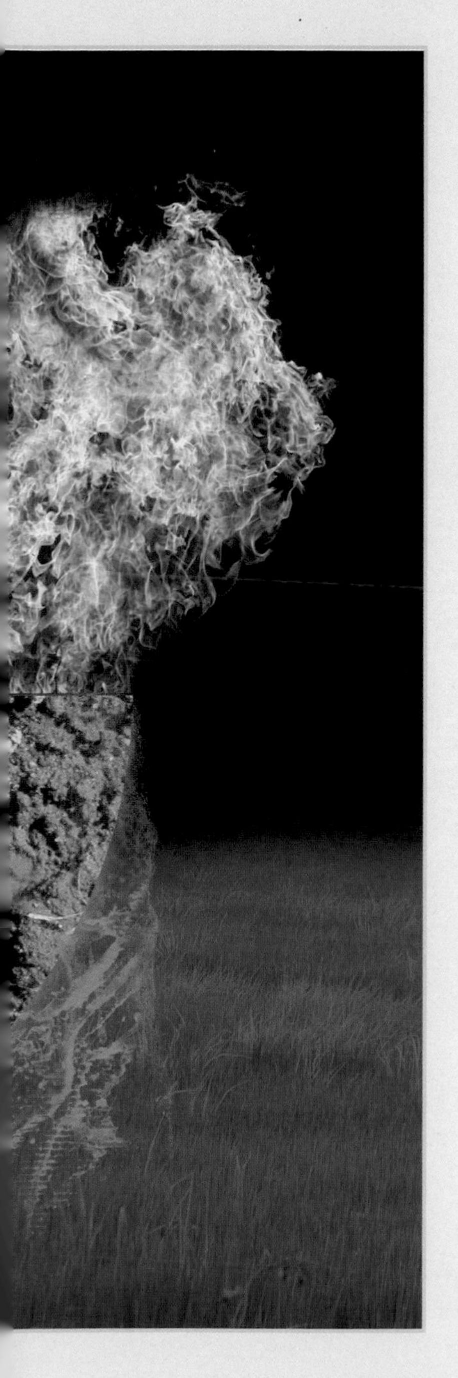

It's difficult to imagine, when looking at a solid object, that it is composed of many very tiny particles and a lot of empty space. It's even stranger when we pause to think that the particles themselves are more space than matter. The idea that matter is not continuous, and even that it contains a lot of empty space—which is a fair description of modern atomic theory—was first suggested around 2,500 years ago. Even so, atomic theory has only been accepted by a majority of scientists for a little over a century. For much of the intervening time, the concept was discredited, and even ridiculed.

From ancient times, humankind has aspired to understand the building blocks of the universe. These fundamental elements, according to the ancient Greeks, were earth, fire, air, and water.

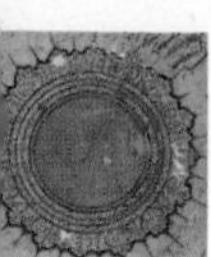

The First Physicist?

The origins of "natural philosophy"—or science, as we call it today—probably lie, as with so much else that underpins western culture, in ancient Athens. The first person we can call by name a physicist is Anaxagoras, who flourished in the fifth century BCE. At a time when logic was in its infancy, he tried to fit his myriad observations and the results of his experimentation into a logical framework that would enable him to understand and explain the nature of the world. Anaxagoras sought a view of the material universe in which superstition or divine intervention need play no part, a scheme in which everything was explicable to the rational mind—a truly scientific model. In limiting himself to types of matter that could be perceived, Anaxagoras set a pattern for physicists for dealing with the visible, physical world that was to last for nearly 2,500 years.

> *"Nothing will come of nothing."*
> *King Lear,* Act I, Scene 1

The Seeds of Matter

For Anaxagoras, the central feature of the natural world was change. He saw everything in constant movement, one thing shifting into another in an endless cycle. Matter, he said, could neither come into existence from nothing nor cease to exist, a belief he shared with earlier thinkers Thales of Miletus and Parmenides (*ca.*515–*ca.*445 BCE). This same belief was presented much later by the French chemist Antoine Lavoisier (1743–94) in the law of the conservation of mass (see page 30). Further, he claimed that all matter was composed of the same fundamental ingredients—essential properties, and perhaps "seeds" of basic substances. The properties always existed in pairs that were polar opposites, such as hot–cold, dark–light, and sweet–sour. There would always be the same quantity of each property in total. The seeds were principally of organic matter (blood, flesh, bark, fur).

Anaxagoras believed that any portion of matter, irrespective of how small it is, contains all possible properties (or materials). This means it must be capable of infinite division. The properties that predominate are evident and give the substance its observable characteristics, while others are latent. So a tree has more bark than fur, but it still has some of each—it just doesn't have enough fur to manifest "furriness." This explains how any substance can be made from any other, as it simply requires taking different proportions of all the properties (or materials) to form the new substance.

Mind Animating Matter

Anaxagoras had an additional ingredient to throw into the melting pot, and this was mind, or *nous*. He did not believe mind to be present in all matter, but only in animate (living or conscious) things. Mind had an additional role, though: at the start of all things, matter was not distinguished into different substances, but a homogenous pile of particles or slurry that became sorted into "proper" matter by the principle of mind.

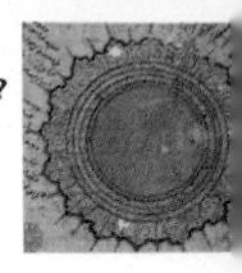

ANAXAGORAS (*ca.*500–*ca.*430 BCE)

Born in Ionia, on the western coast of what is now Turkey, Anaxagoras moved to Athens at the age of 20 where he immediately entered the highest intellectual circles. He became the close companion and instructor of Pericles, political ruler of Athens at the height of the city's power (454–431 BCE). Anaxagoras taught and wrote a treatise on natural philosophy that was later used by the Greek philosopher Socrates (469–399 BCE). His fame spread far and wide, his zeal for the intellectual life and disregard of all fleshly and social pleasures becoming as famous as his teachings. Anaxagoras was so devoted to the life of the mind that he neglected everything else and allowed his substantial inheritance to waste away.

Despite being the leading intellectual figure in Athens, he moved away from the city after about 30 years, and little is known of his later life. He died at Lampsacus on the coast of the Dardanelles aged around 70, but his influence continued for a century after his death.

In Anaxagoras' schema, a natural object such as a badger mixes seeds including fur, blood, and bone with animating nous *or "mind." An inanimate object shares the same seeds in different proportions but has no "mind."*

This sounds horribly like creation by a divine entity, though, and Anaxagoras was adamant that he wanted no superstition or religion in his account of the world. His "mind" was not an intelligent creator but some kind of inspiring element that set in motion the physical forces that whirled elemental matter around, causing it to separate, differentiate, and form bodies such as the Earth and sun. It is difficult to be precise about the role of mind as Anaxagoras' complete text has not survived. However, Plato reports that Socrates bought a copy of Anaxagoras' work because he thought it contained an explanation involving a designing intelligence and was disappointed.

When a tree burns, its constituents are rearranged rather drastically.

All Change

Anaxagoras had a model in which matter could not be created or destroyed, but in which the mutability of the world around us is explained by changing the position of matter over time. If a tree is cut down and the wood made into a boat, the matter has moved and rearranged, but is of the same type and quantity (counting the boat, off-cuts, and sawdust) as before. Other changes require more substantial rearrangement: setting fire to a tree, for example, produces ash, water vapor, and smoke that do not appear to be at all similar to wood. As every object contains, in different proportions, all the possible types of matter and qualities, there is always the potential for each type of matter to be derived from any object—so a plant could grow from soil, for instance, by rearranging or extracting types of matter.

Anaxagoras realized that to make this work, the constituent parts of matter (seeds) must be extremely small, as otherwise the kinds of changes we see every day would not be possible. The requirement for the components of matter to be infinitesimally small was to present insurmountable problems for the model.

Uncuttable Portions

The word "atom" comes from the Classical Greek word "atomos," meaning uncuttable or indivisible. The suggestion that everything is made from very tiny, indivisible particles has its origins in the fifth century BCE with the work of Leucippus and then his pupil Democritus. Far more is known about Democritus (*ca.*460–*ca.*370 BCE) than about Leucippus, to the extent that the Greek philosopher Epicurus

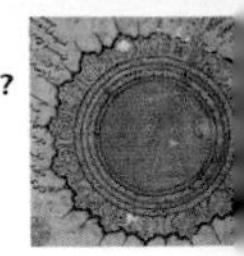

(341–270 BCE) doubted Leucippus even existed. It is impossible to tell what portion of the atomic model came from Leucippus. Atomism holds that the universe comprises matter made from tiny, indivisible particles that exist in a void. The atoms of any particular substance are all the same size and shape and made of the same material.

If atoms are tiny, homogenous (homoiomerous) particles, there is an obvious question—why can't they be further divided? If Democritus had an answer, it has not survived. It may have been that atoms, being homogenous, have no internal void (whereas larger chunks of matter have space between the atoms), and this alone means they cannot be divided.

There is an innate paradox, too, in a model of matter made up of infinitesimal particles. What Anaxagoras meant by infinitesimal was that the particles were smaller than any arbitrarily small measure, but larger than zero. Even so, he believed that every object held an infinite number of particles, as no matter how small a portion he took there would always be some of every type of matter. If atoms or seeds have no extension in space (zero size), then even an infinite number of them could not make up matter of finite size. This dilemma presented insuperable problems for later Greek thinkers, and led the atomic model into the doldrums from which it did not emerge for 2,000 years.

Things and Not-Things

So far, atomism sounds very similar to Anaxagoras' model yet he had all matter floating in the air or *aether* (see page 22), which is a physical substance, while the atomists had particles of matter existing in a void. Democritus (or Leucippus) was the first to postulate a void, though it was clearly needed if matter is to move: in a universe jam-packed full of matter, every bit of space would already be occupied so it could not be occupied by something else moving into it. When something moves, it not only shifts into an empty space or nudges something else into empty space, it also leaves empty space behind it. While earlier thinkers had denied there is a void ("what is not"), Democritus relied on the evidence of our senses—we know that things move—to establish the void as a valid concept. Further, we can see that the universe is made up of many things (it has plurality), whereas if there were no empty space, all matter would be continuous. Plurality and change both require a void.

HOMOIOMERIES

Anaxagoras and later Greek thinkers distinguished between substances that were homoiomerous (homogenous) and those that were not. A homoiomerous substance is one in which all parts are like the whole. So a lump of gold is homoiomerous because no matter how small a chunk of it is taken, it still has the properties of a large chunk of gold. A tree or a ship is not homoiomerous as it can be broken into parts that have different characteristics. To modern eyes, homoiomeries are the elements and pure chemical compounds.

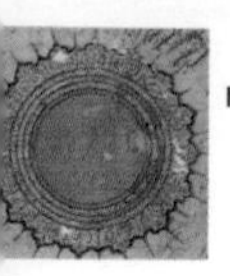

ARISTOTLE (384–322 BCE)

Aristotle was born in Stageira, in Macedonia, to the court physician, but was orphaned at an early age. He moved to Athens at around 18 years of age to study under Plato in his Academy, following advice given him by the Delphic oracle. He became Plato's finest and ultimately most famous pupil. In 342 BCE Aristotle moved back to Macedonia and became tutor to Alexander, the son of Philip II of Macedonia, who was later to become Alexander the Great. Aristotle reviewed the work of all earlier Greek thinkers and then constructed his own views based on the aspects that he thought correct, expanding upon them. He wrote on almost all subjects, including physics. His teachings were preserved by Arab scholars and revived in Europe in Latin translation in the 12th and 13th centuries. Aristotle's scientific ideas dominated western science until the 18th century.

Atomic and Elemental Matter

To the modern mind, atoms and elements are part of the same model of the universe. The elements are the pure chemical substances, each made up of identical atoms, so all of gold is gold atoms, and all of hydrogen is hydrogen atoms. Compounds, on the other hand, contain atoms of two or more elements, so carbon dioxide comprises carbon and oxygen atoms, for instance. In ancient theories of matter, though, atoms and elements belong to different models.

Four—or Five—Elements

Empedocles (*ca*.490–*ca*.430 BCE) taught that everything is made up from four "roots"—earth, air, water, and fire. This model was reworked and championed by Aristotle, perhaps the greatest and most influential thinker in the history of the West. Plato renamed the roots "elements" and Aristotle used this term. Each element is characterized by two properties from the natural contraries—hot–cold and wet–dry. So earth is cold and dry, water is cold and wet, air is hot and wet, and fire is hot and dry. These properties also formed the basis of the model of health and illness based on the four humors proposed by Hippocrates (*ca*.460–*ca*.377 BCE) or his school, which endured into the 19th century.

According to elemental theory, all matter naturally occupied a realm that was associated with its elements and matter is

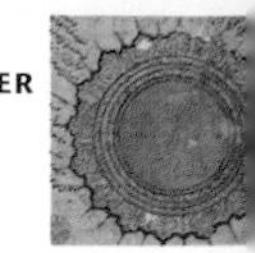

drawn toward its natural realm. Earth occupied the lowest position, fire the highest, with water and air between them. This explained some types of movement in the physical world: heavy objects fall to earth because earth is their principal element; smoke comprises fire and air, which occupy the upper realms, so it rises. Once an element is in its natural place, it will not move unless something causes it to do so.

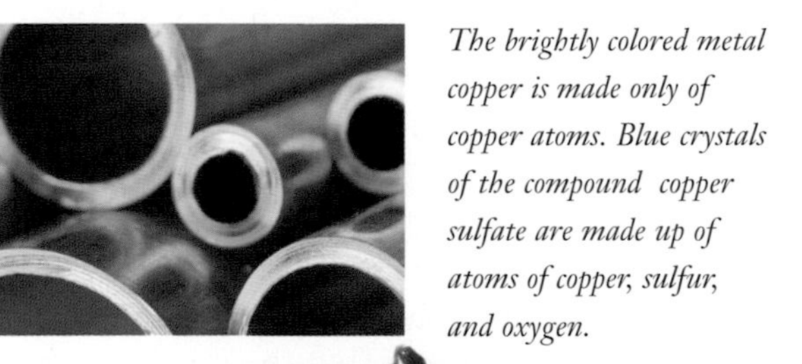

The brightly colored metal copper is made only of copper atoms. Blue crystals of the compound copper sulfate are made up of atoms of copper, sulfur, and oxygen.

In addition to the four elements, there is a very different fifth element (or "quintessence"), called the "aether." The concept of an aether (or "ether") never quite went away, though it went in and out of favor over thousands of years (see page 22).

Although Democritus' atomist model was in fact much closer to reality as understood today, it was the idea favored by Empedocles, Plato, and Aristotle, of a world made up of four elements, that proved most popular. When the Arab thinkers of the early Middle Ages reinvigorated and developed the thinking of Classical Greece, it was this elemental model that was carried forward. From there it was translated into Latin and then other European languages; it remained the cornerstone of thinking about the nature of matter for more than 2,000 years.

CH-CH-CHANGES

While Parmenides could not account for change at all and the atomists used the void to allow matter to change, Aristotle formulated all change as transformation between states. This involved equal "becoming" and "unbecoming"—again a version of the conservation of mass. So to

Allegorical depiction of the four elements in a 12th-century manuscript

become a statue, a block of stone or bronze stopped being a block and became a statue. To become a man, a boy stopped being a child. Each changeable thing already has the potential to be something else, and that potential is realized when it changes. It then loses its potential to become and has "actuality."

THE AETHER: 2,500 YEARS OF AN UNDETECTABLE MEDIUM

The aether, or quintessence, first appears as the fifth element in Ancient Greek thought. This is the element of the heavens and forms no part of earthly matter. It was considered the natural realm of the gods and was unchanging and eternal. It was thought to move only in circles, since a circle is the perfect shape. Differences in densities in the aether were thought to account for the existence of the heavenly bodies. The great French philosopher and mathematician René Descartes (1596–1650) thought vision possible because pressure exerted on the aether was transferred to the eye. The concept of aether or "ether" was revived in the 19th century by the Scottish scientist James Clerk Maxwell (1831–79) to explain the transport of light and other forms of electromagnetic radiation.

The Dutch physicist Hendrik Lorentz (1853–1928) developed a theory of an abstract electromagnetic medium in 1892–1906, but when Albert Einstein published his special theory of relativity in 1905, he dispensed with the aether altogether.

More recently, a number of cosmologists have again proposed some kind of aether suffusing the cosmos, perhaps linked with dark matter.

Indian Atomism

The Greeks were not the only thinkers to come up with a type of atomic theory. Indian philosophers also suggested that matter may be made up of tiny particles. It is unclear whether the Greeks or the Indians derived it first, and whether they developed it independently or whether one tradition influenced the other. The Indian philosopher Kanada may have lived in either the sixth or second century BCE (historians cannot agree). If the earlier date is correct, Kanada's atomism predates the Greek tradition, and may have influenced it.

Kanada's theory of atoms complemented the elemental theory in that he proposed five different types of atom, one for each of the five elements that made up the Indian model of matter—fire, water, earth, air, and aether, the same as in Aristotle's model. Atoms—or *parmanu*—are attracted to each other, and will group together. A diatomic particle, *dwinuka*, has properties belonging to each component; these then group into triatomic clusters that were thought to be the smallest visible components of matter. The variety and different properties of matter are accounted for by the differing combinations and proportions of the five types of *parmanu*. In the version of Kanada's atomism developed by the Vaisesika school, atoms could have a combination of 24 possible properties. Chemical and physical changes in matter come about when *parmanu* recombine.

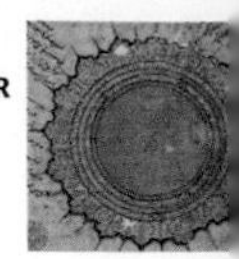

KANADA (KASHYAPA)

The Indian philosopher Kanada was born in Gujarat, India. According to tradition, he was originally named Kashyapa but as a boy was given the name Kanada (from Kana, meaning grain) by the sage Muni Somasharma on account of his fascination with tiny things. His main area of study was a type of alchemy (see page 26). He proposed an atomic theory of matter, which reportedly came to him as he was walking along eating and throwing away small particles of food. He is said to have realized that he could not continue to divide the food into ever smaller pieces, but ultimately it must be made of indivisible atoms.

Unlike the Greek philosophers, Kanada believed atoms could come into being or cease to exist instantly, but could not be destroyed by physical or chemical means.

The Jain theory of atomism dates to the first century BCE or earlier. It sees the whole world, with the exception of souls, being composed of atoms, each of which has one kind of taste, one smell, one color, and two kinds of touch characteristic. Jain atoms were in constant motion, usually in straight lines, though they could follow a curved path if attracted to other atoms. There was also a concept of a polar charge, with particles having a smooth or rough characteristic that enabled them to bind together. Atoms could combine to produce any of six "aggregates"—earth, water, shadow, sense objects, karmic matter, and unfit matter. There were complex theories of how atoms behaved, reacted, and combined.

Islamic Atomism

Whether Indian or Greek theories were earliest, both were brought together by early Islamic scholars. The teachings of the Ancient Greeks survived in the Eastern (Byzantine) Roman Empire and were resurrected by early Arab scholars who translated and commented on them. There were two principal forms of Islamic atomism, one closer to Indian and one to Aristotelian thought. The most successful was the Asharite work of al-Ghazali (1058–1111). For al-Ghazali, atoms are the only material things that are eternal; everything else lasts only for an instant and is said to be "accidental." Accidental things can't be the cause of anything except perception.

Al-Ghazali was an Asharite—a sect that believed human reason could not establish truths about the physical world without divine revelation.

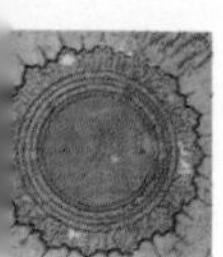

A few years later, the Spanish-born Islamic philosopher Averroes (Ibn Rushd, 1126–98) rejected al-Ghazali's model and commented extensively on Aristotle. Averroes was very influential on later medieval thought and was instrumental in Aristotle being absorbed into Christian and Jewish scholarship.

Much Arab work was translated into Latin in the early Middle Ages, introducing Classical Greek thought into Western Europe. Aristotle's teachings were adopted by the Catholic Church wherever they did not directly contradict the Bible or influential Christian thinkers. By this route, they formed the foundation of the accepted scientific and philosophical models that were current in the West until the Renaissance, when European thinkers finally began to challenge and check the teachings of the ancients.

From Atoms to Corpuscles

In the 13th century, an anonymous alchemist known as Pseudo-Geber set forth a theory of matter based on tiny particles, which he called "corpuscles." (Pseudo-Geber's odd name comes from him signing his works Geber, which was the Latinized form of the name Jabir ibn Hayyan, an eighth-century Islamic alchemist, even though the texts were not actually translations of Geber's works.) Pseudo-Geber proposed that all physical materials have an inner and outer layer of corpuscles. He believed that all metals were made from corpuscles of mercury and sulfur in different proportions. He used this belief in support of alchemy (see box, page 26) since it meant all metals had the necessary ingredients for becoming gold—they just needed the appropriate refining or rearrangement.

Something similar to Pseudo-Geber's view was described by Nicholas of Autrecourt (*ca.*1298–*ca.*1369). Autrecourt took up the debate raging in Paris, the intellectual center of Europe at the time, regarding the divisibility or indivisibility of a continuum. This question arose from Aristotle's statement that a continuum cannot be made up of indivisible particles. He believed that all matter, space, and time were made up of atoms, points, and instants and that all change is the result of rearranging atoms. Various of Autrecourt's

Imaginary debate between the Aristotelian Averroes (left) and the neo-Platonic philosopher Porphyry, who died 800 years before Averroes' birth.

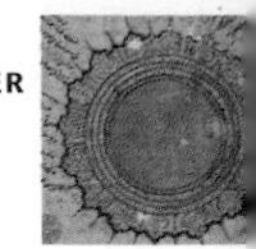

views offended the Church and he had to retract them after being put on trial in 1340–46. For him, all movement was inherent in the moving object (as movement is reduced to the motion of particles). His view that time as well as matter is granular, being made up of discrete instants, was not picked up by later thinkers.

Pierre Gassendi was a proponent of corpuscularianism.

A variant of early atomism became popular in the 17th century and was supported by the Irish chemist Robert Boyle, the French philosopher Pierre Gassendi (1592–1655), and Isaac Newton among others. Known as "corpuscularianism," it differed from atomism in that corpuscles need not be indivisible. Indeed, proponents of alchemy (including Newton) used the divisibility of corpuscles to explain how mercury could insinuate itself between the particles of other metals, paving the way for its transmutation into gold. The corpuscularians held that our perceptions and experiences of the world around us result from the actions of the tiny particles of matter on our sense organs.

From Corpuscles Back to Atoms

Atomism was not truly revived until Pierre Gassendi proposed a skeptical view of the world in which everything that happened did so because of the movement and interaction of minute particles following natural laws. Gassendi excluded thinking beings from his scheme, but in other respects the theory he published in 1649 was astonishingly accurate. He thought the properties of matter were produced by the shapes of atoms, that atoms could join together into molecules, and that they existed in a vast void—so that most of matter was actually non-matter. Gassendi's insight was not as influential as it should have been because the much more influential Descartes was directly opposed to it, flatly denying that there could be a void. In one regard, though, Gassendi and Descartes were in agreement: both believed the world was essentially mechanistic and followed laws of nature.

Robert Boyle brought atomism to the fore again a few years after Gassendi's death. In 1661, he published *The Sceptical Chymist*, describing a universe made up entirely of atoms and clusters of atoms, all in perpetual motion. Boyle proposed that all phenomena are the result of collisions between atoms in motion, and called for chemists to investigate elements as he suspected there were more than the four Aristotle had identified.

The Age of Reason

The Age of Reason is the name generally given to the period starting around 1600 when the philosophical mood of Western Europe and the new colonies in America was one of confidence in human endeavor.

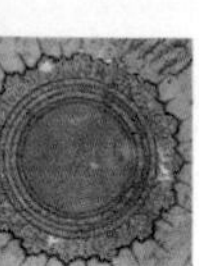

It continued the flourishing of optimism and achievement begun in the Renaissance, and completed the shift from the denigrating or humble view of humankind as flawed sinners that predominated in the Middle Ages to a view that celebrated human achievement and potential. The Age of Reason both drove and was driven by developments in science, technology, philosophy, political thought, and the arts.

The philosophy of the period is sometimes divided into two camps, rationalist and empiricist. The rationalists held that reason was the route to knowledge, while the empiricists favored observation of the world around us. This roughly followed the division between Plato (rationalist) and Aristotle (empiricist) in ancient thought. The empiricist view led directly to scientific experimentation and observation, while rationalism favored mathematical and philosophical approaches. There is no clear division between the two, though, as conclusions reached by rational deduction are often amenable to testing by empirical methods. Together these approaches formed the foundation of the scientific revolution. The development of the scientific method, one of the triumphs of the Age of Reason, changed the course of scientific discovery forever.

The Birth of Solid-State Physics

Accepting that matter is made up of tiny particles, whether we call them atoms or corpuscles, led to obvious questions such as what shape do they have? How do they join

ALCHEMY

The best-known aims of the philosophical and scientific endeavor of alchemy are to change base metals into gold, through transmutation, and to produce an elixir of life. The fabled philosopher's stone was often thought to be an essential component of the elixir of life, or of the transmutation process, or both. Alchemy has been practiced in various forms in Ancient Egypt, Mesopotamia, Ancient Greece, China, and the Islamic Middle East, as well as in the European Middle Ages and Renaissance. Alchemy lies behind modern chemistry and pharmacology, and in Chinese alchemy the production of medicines was a major activity. Attempts at transmutation often began with lead, but other base metals could be used. Needless to say, none of the alchemists' methods worked.

Alchemists worked at transmutation and distillation in early laboratories.

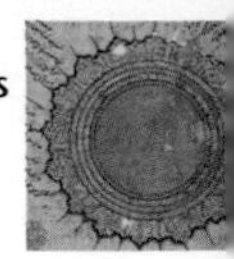

THE POWER OF NOTHING

The German scientist Otto von Guericke (1602–86) invented—or discovered—nothing. Literally. He proved that a vacuum could exist, which previous scientists had denied. After experimenting with bellows and developing an air pump, he put on a spectacular demonstration in front of Emperor Ferdinand III in 1654. He constructed metal spheres from two hemispheres and pumped out the air. He then showed the power of the vacuum—or rather the power of atmospheric pressure—by demonstrating that not even two horses could pull the hemispheres apart.

Two years before his death in 1689, Robert Boyle was already in poor health.

together into contiguous matter? How do different types of matter react and interact? How do physical changes (melting, freezing, sublimation) relate to the model of particles? Physicists of the 17th century deduced models of the structure of matter from observing the properties and behavior of substances—which sometimes led them to pretty bizarre deductions.

After watching wrought iron being produced, Descartes concluded that the particles of iron were somehow joined together into grains, and that cohesion within the grains was greater than cohesion between the grains. He failed to notice that the "grains" in wrought iron form a crystalline structure, though. Although in theory microscopes could have revealed

"[Robert Boyle] is very tall (about six foot high) and straight, very temperate. and virtuous and frugal: a bachelor; keeps a coach; sojourns with his sister, the Lady Ranelagh. His greatest delight is chemistry. He has at his sister's a noble laboratory and several servants (apprentices to him) to look after it. He is charitable to ingenious men that are in want, and foreign chemists have had large proof of his bounty, for he will not spare for cost to get any rare secret. At his own cost and charges he got translated and printed the New Testament in Arabic, to send into the Mahometan countries. He was not only a high renown in England, but abroad; and when foreigners come hither, 'tis one of their curiosities to make him a visit."

John Aubrey, *Brief Lives*

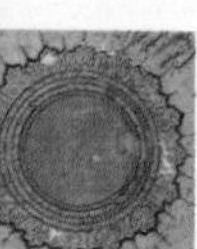

> *"There are therefore Agents in Nature able to make the Particles of Bodies stick together by very strong Attractions. And it is the Business of experimental Philosophy to find them out. Now the smallest Particles of Matter may cohere by the strongest Attractions, and compose bigger Particles of weaker Virtue and many of these may cohere and compose bigger Particles whose Virtue is still weaker, and so on for divers Successions, until the Progression end in the biggest Particles on which the Operations in Chymistry, and the Colours of natural Bodies depend, which by cohering compose Bodies of a sensible Magnitude. If the Body is compact, and bends or yields inward to Pression without any sliding of its Parts, it is hard and elastick, returning to its Figure with Force arising from the mutual Attraction of its Parts. If the Parts slide upon one another, the Body is malleable or soft. If they slip easily, and are of a fit Size to be agitated by Heat, and the Heat is big enough to keep them in Agitation, the Body is fluid..."*
>
> Isaac Newton, notes to the second edition of *Opticks*, London 1718

such structures, they were not in common use until the second half of the 17th century; even then they were used mostly in biological studies. Of course, no microscope could show the shape of atoms or molecules.

The Cartesian physicist Jacques Rohault (1618–72) suggested in 1671 that plastic (or pliable) materials had particles with complicated textures that are tangled together, while brittle materials have particles with a simple texture that touch one another at only a few points. In 1722 the French thinker René Antoine Ferchault de Réaumur (1683–1757) determined that, contrary to previous belief, steel is not purified iron, but iron to which "sulfurs and salts" have been added and that particles of these substances lie between the iron particles.

With no other method than imagination on which to rely, physicists came up with some outlandish suggestions for the shapes of particles. Nicolaas Hartsoeker (1656–1725) claimed in 1696 that air is made up of hollow balls constructed of wire-like rings, that mercuric chloride is a ball of mercury stuck with needle- or blade-like spikes of salt and vitriol, and that iron has particles with

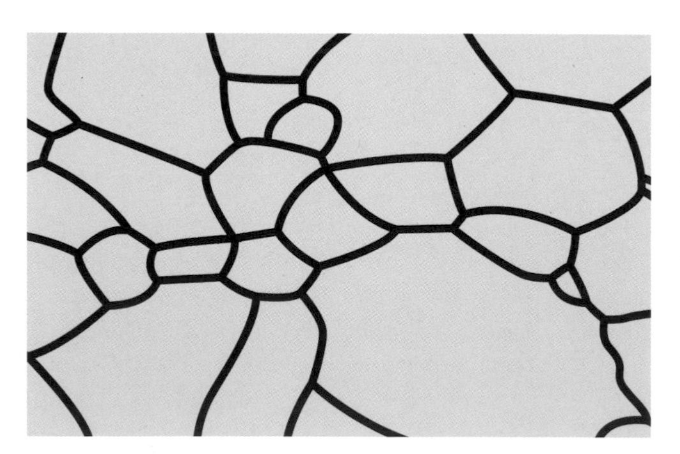

The microstructure of steel: scientists of the 17th century did not look at metal with microscopes.

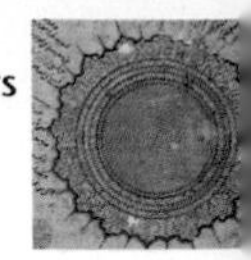

Descartes watched iron being smelted, a process shown in this contemporary image.

teeth that interlock to make it hard when cold. Iron is malleable when heated, he argued, as the particles separate sufficiently to allow them to slide over one another. Thinking up the structures of matter was a game, and Hartsoeker finished by encouraging his readers to join in: "I do not wish to deprive the reader of the pleasure of himself making the search following the principles that have been established above."

Atoms and Elements

Robert Boyle was right to encourage chemists to look for more elements than earth, water, air, and fire but it was some time before a table of chemical elements was formulated. Antoine Lavoisier produced the first modern work on chemistry in 1789 and he included in it a list of 33 elements—substances that could not be broken down further. Unfortunately, Lavoisier's list included light and "caloric," which he thought was a fluid that produced loss or gain of heat through its motion (see page 89). Lavoisier did

> *"Soul of the World! Inspir'd by thee,*
> *The jarring Seeds of Matter did agree,*
> *Thou didst the scatter'd Atoms bind,*
> *Which, by thy Laws of true proportion join'd,*
> *Made up of various Parts one perfect Harmony."*
>
> Nicholas Brady, "Ode to St Cecilia," *ca.*1691

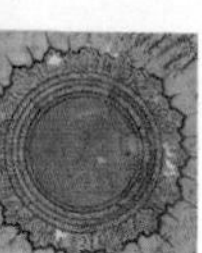

ANTOINE-LAURENT DE LAVOISIER (1743–94)

Antoine Lavoisier (as he was styled after the French Revolution, when a fancy, noble name became a liability) was the son of a wealthy lawyer and had originally trained in the law himself. He turned to science instead, first studying geology but later becoming increasingly interested in chemistry. He had his own laboratory, and it and his house soon became a magnet for free-thinkers and scientists.

Lavoisier has been called the father of modern chemistry. His achievements were considerable and varied. As well as listing the elements, he recognized the role of oxygen in combustion and respiration, and that similar reactions were involved in each. This overturned the popular and ancient theory of phlogiston (a substance supposedly released when matter is burned—see page 86).

Lavoisier was politically liberal and supported the ideals that led to the French Revolution. He served on a committee that proposed economic reform, and suggested improvements to the dire conditions in Parisian jails and hospitals, but this did not ultimately save him. He was executed by guillotine during the Terror in 1794. It is said that he asked to delay the execution so that he could finish his experiments, but was told, "The Republic has no need of scientists." A story that he asked an assistant to count how many times he continued to blink after his head was cut from his body is widely reported but probably apocryphal.

Antoine Lavoisier, the first true chemist

not consider his list of elements exhaustive, leaving the door open for further investigation and later discoveries. Nor did he organize his list of elements into the periodic table—that work was left for the Russian chemist Dmitri Mendeleev (1834–1907) to complete in 1869. The periodic table is relevant for the history of physics in that organizing the elements according to their properties revealed the significance of atomic number and its relation to valency—the way elements bond together.

As an empiricist, Lavoisier claimed that in his work he had "tried... to arrive at the truth by linking up facts; to suppress as much as possible the use of reasoning, which is often an unreliable instrument which deceives us, in order to follow as much as possible the torch of observation and of experiment." Another contribution that was later to prove important in understanding chemical reactions at the atomic level was Lavoisier's law of conservation of mass—the recognition that mass is never lost or gained in the process of a chemical reaction. But despite coming up with a list of elements, he did not believe in atoms, which he considered philosophically impossible.

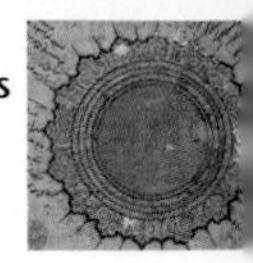

All in Proportion

> *"It required only a moment to sever that head, and perhaps a century will not be sufficient to produce another like it."*
> Mathematician and astronomer Joseph-Louis Lagrange on the execution of Lavoisier, 1794

Deciding that atoms exist is a good start, but in order to build continuous matter from them, and in more varieties than just the elements identified by Lavoisier, there needed to be some means of joining atoms together. Exactly how atoms adhere into groups was a puzzle to early atomists. Newton wrote of "Agents in Nature" that could hold atoms together.

The first step in investigating how atoms combine was to determine the ratios in which they join in compounds. The French chemist Joseph Proust (1754–1826) deduced the law of definite proportion from experiments he carried out between 1798 and 1804 while director of the Royal Laboratory in Madrid. His law states that in any particular chemical compound, the elements always combine in the same whole-number ratios by mass.

Just a few years after Lavoisier was guillotined in Paris, the English chemist John Dalton (1766–1844) developed this idea further and laid the foundations of modern atomic theory. In work that he began in 1803 and published in 1808, he set out five observations about atoms:

- All elements are made of atoms.
- All atoms of a given element are identical.
- Atoms of one element differ from atoms of every other element and they can be distinguished by their atomic weights.
- Atoms cannot be created, destroyed, or divided by chemical processes.
- Atoms of one element can combine with atoms of another to make a chemical compound; a given compound always contains the same proportion of each element.

Dalton developed the law of multiple proportions. Instead of just looking at a single compound formed by two elements, he looked at elements that can combine in more than one way. He found that the relative proportions are always small whole-number ratios. So, for example, carbon and oxygen can form carbon monoxide (CO) or carbon dioxide (CO_2). Using the weights of combining oxygen and carbon, in CO the

Otto von Guericke conducted experiments to demonstrate a vacuum.

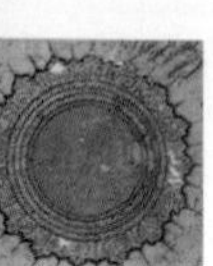

proportion is 12:16 and in CO_2 is 12:32. So the ratio of oxygen in CO to that in CO_2 is 1:2.

From the ratios in which masses of elements combine, it was possible to work out relative atomic masses. Dalton calculated atomic mass according to the mass of each element in a compound, using hydrogen as his base unit (1). However, he incorrectly assumed that simple compounds are always formed in the ratio 1:1—so he thought water was HO and not H_2O—and as a result made some serious errors in his table of atomic numbers. Dalton was also unaware that some elements exist as diatomic molecules (that is in pairs, such as O_2). These basic errors were corrected in 1811 when the Italian chemist Amedeo Avogadro (1776–1856) realized that a fixed volume of any gas at the same temperature and pressure contains the same number of molecules (related to Avogadro's Constant, 6.0221415×10^{23} mol^{-1}). From this Avogadro calculated that as two liters of hydrogen react with one liter of oxygen, the gases combine in the ratio 2:1. Avogadro—full name Lorenzo Romano Amedeo Carlo Bernadette Avogadro di Quaregna e Cerreto—is now considered the originator of atomic-molecular theory.

Atoms—True or False?

Although Dalton's work looks convincing with hindsight, the scientists of the day were not bowled over by his explanation and physicists remained divided into two camps—those who accepted the likely existence of atoms and those who did not. Luckily, there were good practical reasons to continue looking at gases. The development of the steam engine led to a growing interest in thermodynamics and so to some consideration of the properties and behavior of atoms. The behavior of atoms could be related to the action of hot gases on a much larger scale, and so to the laws of thermodynamics that emerged in the middle of the 19th century.

The first visual evidence that matter is made up of tiny particles was discovered—though not immediately explained—in 1827 by the Scottish botanist Robert Brown (1773–1858). While examining minuscule pollen grains in water under the microscope, Brown noticed that they moved around constantly as though something invisible was bumping into them. He found the same movement occurred even when using pollen grains that had been stored for a hundred years, so demonstrating that the movement was not initiated by the living grains themselves. Brown couldn't explain what he saw, so what is now called Brownian motion attracted little attention for a long time. In 1877 J. Desaulx revisited it, suggesting "In my way of thinking the phenomenon is a result of thermal molecular motion in the liquid environment (of the particles)." The French physicist Louis Georges Gouy (1854–1926) found in 1889 that the smaller the particle, the more pronounced the movement, which was clearly in line with Desaulx's hypothesis. The Austrian geophysicist Felix Maria Exner (1876–1930) measured the motion in 1900, relating it to particle size and temperature. This paved the way for Albert Einstein to create a mathematical model to explain Brownian motion in 1905. Einstein was certain that molecules were

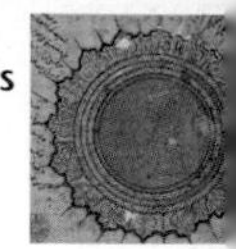

responsible for the movement, and arrived at the first estimates for the size of molecules. The theory was validated by the French physicist Jean Perrin (1870–1942) in 1908 when he measured the size of a water molecule using Einstein's model. This was the first experimental evidence for the existence of molecules, for which Perrin was awarded the Nobel Prize for Physics in 1926. At last, it could only be a particularly truculent scientist who could deny the existence of atoms and molecules.

Are Atoms Divisible?

If we take Democritus' view that atoms are the tiniest indivisible components of matter, then atoms are not, strictly speaking, atoms. Even as Einstein and Perrin were proving the existence of atoms, evidence for smaller—subatomic—particles was starting to emerge. With British physicist Joseph John (J.J.) Thomson's discovery of the electron in 1897, the indivisibility of the atom was about to be challenged. The atom would enjoy its title of ultimate particle for only a few brief years. But before we delve inside the atom, we will look at some phenomena not usually considered to be made of anything: light, forces, fields, and energy.

ATOMS: A MATTER OF LIFE AND DEATH

Arguments about whether atoms did or did not exist raged throughout the 19th century, with some physicists claiming atoms were only a useful mathematical construct and not a part of reality. The dispute caused the emotionally and mentally fragile Ludwig Boltzmann (1844–1906), an Austrian physicist and confirmed atomist, to seek a philosophy that could accommodate both views and put an end to the arguments. He borrowed a notion from the German physicist Heinrich Hertz (1857–94) that suggested atoms were *"Bilder"* (pictures). This meant that atomists could think of them as real and anti-atomists could think of them as an analogy or image. Neither side was satisfied. Boltzmann decided to become a philosopher to find a way of refuting the arguments against atomism. At a 1904 physics conference in St Louis, United States, Boltzmann found most physicists to be anti-atom, and he was not even invited to attend the physics section. In 1905 he started corresponding with the German philosopher Franz Brentano (1838–1917) in the hope of demonstrating that philosophy should be expunged from science (a view echoed by the British cosmologist Stephen Hawking in 2010), but became discouraged. Disillusionment with the majority of physicists who rejected atomism eventually contributed to Boltzmann's suicide by hanging in 1906.

Ludwig Boltzmann

CHAPTER 2

Making Light Work—OPTICS

Human beings have exploited light from the sun, moon, and stars, from fires, and later from lamps for millennia. Light is so essential to our existence that it has often been tied up with religious and superstitious beliefs as a life-giving or creating force. So, for most of recorded history, light has occupied a special place. Over the centuries, people have thought of it as a deity, an element, a particle, a wave, and finally as a wave-particle. Because light is so intricately bound up with sight, the study of optics has included light and vision together. It was not until around a hundred years ago that scientists began to recognize that visible light was only one part of a whole spectrum of electromagnetic radiation.

The discovery that white light comprises light of different colors was a breakthrough in the study of optics.

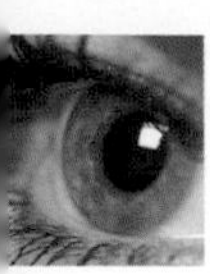

A First Look at Light

Ideas about the nature of light were first recorded in fifth- or sixth-century India. The Samkhya school considered light to be one of the five fundamental "subtle" elements from which the "gross" elements are constructed. The Vaisheshika school, which took an atomist view of the world, held that light was made up of a stream of fast-moving fire atoms—not so very different from the current concept of the photon. The first-century-BCE Indian text *Vishnu Purana* referred to sunlight as the "seven rays of the sun."

The ancients could not disentangle light from vision. In the sixth century BCE, the Greek philosopher Pythagoras suggested that beams travel from the eye like feelers and that we see an object when the beams touch it, a model called emission (or extramission) theory. Plato also believed that rays emitted by the eyes made vision possible, and Empedocles, writing in the fifth century BCE, spoke of a fire that shone out of the eye. This view of the eye as a type of torch couldn't explain why we can't see as well in the dark as in daylight, though, so Empedocles suggested that these beams from the eye must interact with light from another source, such as the sun or a lamp.

The earliest surviving work on optics is by the Greek thinker Euclid (330–270 BCE), who also accepted the emission model. Better known as a mathematician, Euclid started the study of geometric optics, writing on the mathematics of perspective. He related the size of an object to its distance from the eye, and stated the law of reflection: that the angle of incidence is equal to the angle of reflection so that the reflected image seems to be as far behind the mirror as the object is in front of it.

Title page of De rerum natura (On the nature of things) *by Lucretius*

> *"The light and heat of the sun; these are composed of minute atoms which, when they are shoved off, lose no time in shooting right across the interspace of air in the direction imparted by the shove."*
>
> Lucretius, *On the nature of the Universe,* 55 CE

Around 300 years later, another innovative Greek mathematician, Hero of Alexandria (*ca.*10–70 CE), showed that light always

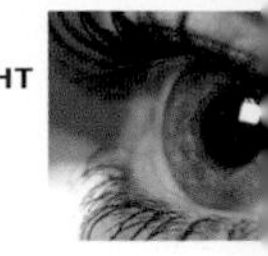

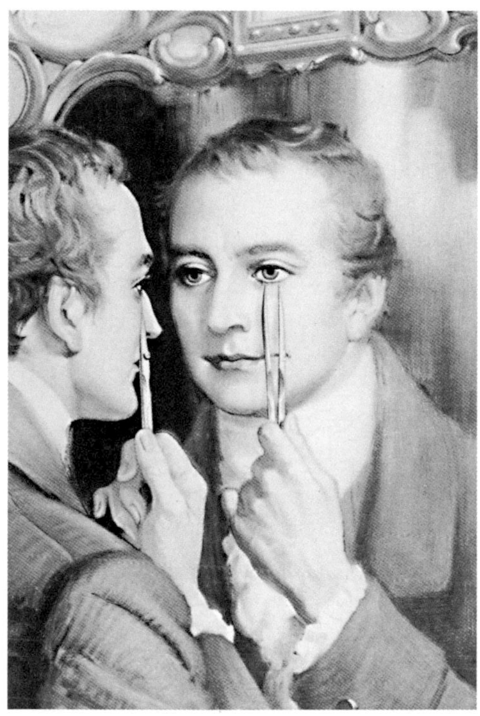

The angle of incidence is equal to the angle of reflection, so Thomas Young's reflection appears to be as far behind the mirror as he is in front of it.

follows the shortest possible path as long as it is traveling through the same medium. If light is both propagated and observed in air, for instance, there will be no deflection. He realized that reflecting the light from flat mirrors did not affect this principle and again demonstrated that the angle of incidence and reflection are equal.

Playing with Light

As Classical Greece declined as the cultural center of Europe, much intellectual endeavor, including the burgeoning physical sciences, also declined. The few remaining Greek thinkers moved eastward. The earliest experimental work on light was carried out by the Greek astronomer Claudius Ptolemy (*ca.*90–*ca.*168 CE) while working in the Library of Alexandria in Roman Egypt. He found that on entering a denser medium (such as going from air to water), light bends in a direction perpendicular to the boundary. He accounted for this by suggesting that light slows down as it enters the denser medium.

Although Ptolemy accepted the emission model of vision, he concluded that the rays *from the eyes* behaved in the same way as the rays of light that traveled *to the eyes*, and so he finally brought the theories of vision and light together. It would be many centuries, though, before it was accepted that vision was entirely the result of light falling on the eye and that the eye does not in any way "reach out and grab" images of the surrounding world. That all-important step was taken around 1025 by the Arab scholar Ibn al-Hassan ibn al-Haytham, who was known as Alhazen in Europe. His work was translated into Latin as *De aspectibus (On perspective)* and was highly influential in medieval Europe. Al-Haytham built on the

Euclid, the Greek mathematician

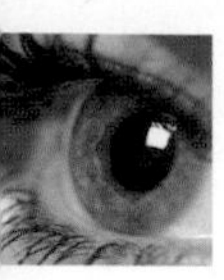

Refraction causes an object partly in water and partly in air to appear disjointed or bent at the boundary between media.

work of the earliest Arab scientist to work on optics, al-Kindi (*ca.*800–70) who proposed "that everything in the world... emits rays in every direction, which fill the whole world." Al-Haytham asserted that rays communicating light and color came from the external world to the eye. He described the structure of the eye and how lenses work, made parabolic mirrors, and gave values for the refraction of light. Al-Haytham also stated that the speed of light must be finite, but it was another Arab scientist, Abu Rayhan al-Biruni (973–1048), who first discovered that the speed of light is considerably greater than the speed of sound.

Ibn al-Haytham

Al-Haytham's work was extended by Qutb al-Din al-Shirazi (1236–1311) and his student Kamal al-Din al-Farisi (1267–1319), who explained how a rainbow is created by splitting white sunlight into the constituent colors of the spectrum. At about the same time, the German professor Theodoric of Freiburg (1250–1310) used a spherical flask of water to show that a rainbow is created as sunlight is refracted on passing from air into a water droplet, is then reflected within the water droplet, and refracted again—passing back from water to air. He correctly gave the angle of the rainbow (between the center and the halo) as 42 degrees. Even so he could not work out what caused a secondary rainbow. It was René Descartes who found, 300 years later, that it is a second reflection of the light inside the water droplets that gives rise to the secondary rainbow and also causes the colors to be reversed.

God's Light

The writings of the Arab scientists were translated into Latin, often by scholars working in Moorish (Arab-controlled) Spain, and soon spread around Europe. The work on optics was picked up by early European scientists, including Englishmen Richard Grosseteste (*ca.*1175–1253) and later the English scholar

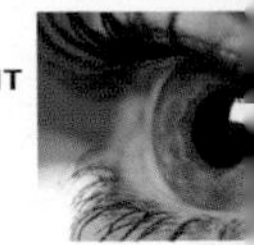

A rainbow is produced by refraction and reflection as light shines into water droplets.

Roger Bacon (*ca.*1214–1294). Grosseteste was working at a time when the heavy reliance on Plato was giving way, under the re-emergence of Aristotle's works from the Arab tradition. He drew on Aristotle, Averroes, and Avicenna in constructing his own work on light. As a bishop, Grosseteste took as his starting point God's

IBN AL-HAYTHAM (965–1040; ALSO ALHAZEN)

Born in Basra, then part of the Persian Empire, al-Haytham trained in theology and tried to resolve differences between the Sunnah and Shi'ah sects of Islam. Failing in that, he turned instead to mathematics and optics. Most of his work on optics was carried out in a period of 10 years while imprisoned in Cairo, having been labeled mad. It seems that he feigned insanity after an over-ambitious claim to be able to stem the flooding of the Nile with one of his engineering projects brought him into trouble. In order to test his hypothesis that light does not bend in air, al-Haytham made the first known camera obscura—a box with a hole at one end to let in light and form an image on the opposite surface that can be traced on to paper. He was a firm believer in using experiment to test his theories. As a rigorous experimental physicist, he is sometimes credited with inventing the scientific method.

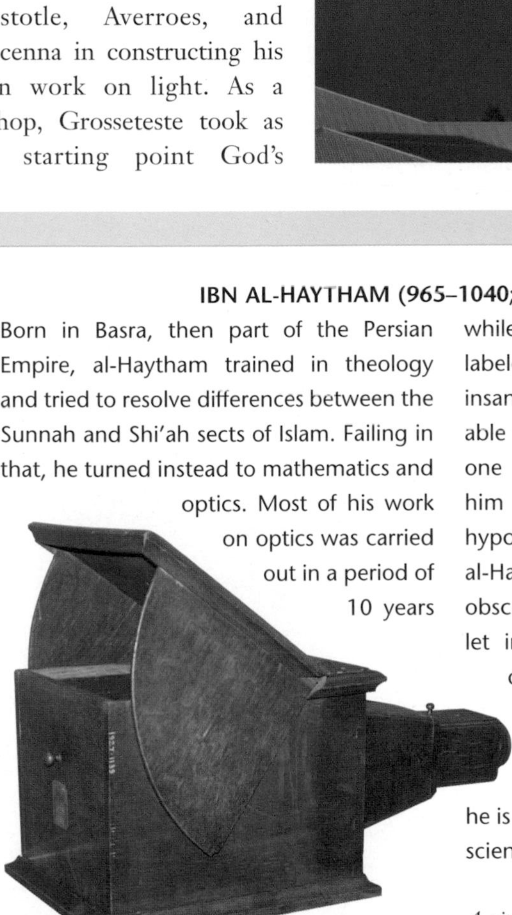

A pinhole camera, or camera obscura

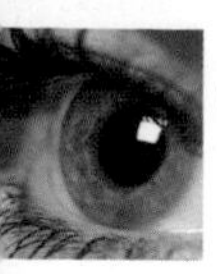

> *"The seeker after truth is not one who studies the writings of the ancients and, following his natural disposition, puts his trust in them, but rather the one who suspects his faith in them and questions what he gathers from them, the one who submits to argument and demonstration."*
>
> Ibn al-Haytham

creation of light in Genesis 1.3: "Let there be light." He saw the process of creation as a physical process driven by expanding and contracting concentric spheres of light. Light is, he argued, infinitely self-generating, as a sphere of light grows instantly from a single point light source. His work is more metaphysics than physics, and is highly original in postulating a method of creation based on the action of light as "first form." An interesting aside and further testament to Grosseteste's originality is that he seems to be the first western thinker to have suggested multiple infinities: "... the sum of all numbers, odd and even, is infinite, and thus is larger than the sum of all even numbers, even though this too is infinite; for it exceeds it by the sum of all the odd numbers."

Roger Bacon, who moved from the University of Oxford to that of Paris, mastered most of the Greek and Islamic texts on optics between 1247 and 1267 and produced his own text, *Optics*. He later set out a program of study that included sciences not then taught at university, and a

ARISTOTLE, IN AND OUT OF FAVOR

The rediscovery of Aristotle's work in Europe, through Latin translations of the texts preserved by Arab scholars, was not immediately popular with the Roman Catholic Church. Aristotle's *Libri naturales* (books of natural science) were condemned by the University of Paris in 1210 and again in 1215 and 1231, which meant they could not be taught. But by around 1230 all of Aristotle's works were available in Latin, so the Paris faculty gave up the fight and in 1255 Aristotle was back on the syllabus and compulsory reading. Roger Bacon, working in Paris at the time, was one of the first to see the effect of letting Parisian scholars loose in Aristotle's playground.

Medieval manuscript copy of Aristotle's Physics *in Latin translation*

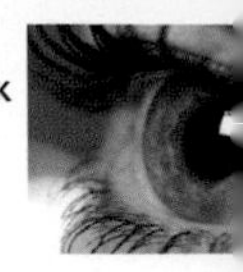

model of experimental science based on his work in optics. He suggested that knowledge of linguistics and science could further and support study in theology, perhaps in an attempt to mollify the Roman Catholic Church. However, the stranglehold of the Church continued to stifle scientific developments for many centuries, with Catholic authorities silencing and even executing scientists who spoke out against the received Biblical version of physical events and phenomena.

Out of the Dark

Truly important original work on optics and light in Europe did not appear until the Renaissance. The towering figures of 16th- and 17th-century science, such as Nicholaus Copernicus (1473–1543), Galileo Galilei (1564–1642), Johannes Kepler (1571–1630), and Isaac Newton (1642–1727), finally dismantled the Aristotelian model of the universe that had dominated scientific thought for nearly 2,000 years and laid out the laws of mechanics and optics that would remain unchallenged for another four or five centuries. Of these, Kepler and Newton were the most important for optics.

Kepler, a German mathematician and astronomer, believed that God had constructed the universe according to an

Hungarian stamps commemorating Kepler and his contribution to space science

GALILEO'S TELESCOPE

Galileo was in Venice when he heard of the development of the telescope; a visiting Dutchman had come to Italy to sell the instrument to the Venetian senate. Desperate to beat him to it,

Galileo presents his telescope to Leonardo Donato, Doge of Venice, in 1609.

Galileo built a telescope in only 24 hours that was better than any in existence. Instead of using two concave lenses, which produced an inverted image, Galileo's telescope had one convex and one concave lens and produced an image the right way up. The senate was persuaded to defer the decision on buying the Dutch telescope. Galileo then produced an even better one that he presented to the Doge of Venice, securing the victory and tenure in his professorial post at Padua University.

intelligible plan and that its workings were therefore discoverable through the application of scientific observation and reasoning. Although more famous for his extensive work in astronomy, Kepler introduced the technique of tracing rays of light point-to-point to determine and explain their path. From this, he deduced that the human eye works by refracting light rays that enter through the pupil and focusing them on to the retina. He explained how spectacle lenses work—they had been in use for around 300 years, but no one really understood the principles behind them—and when telescopes became more widely used, around 1608, he explained how they worked, too. Kepler published his work on optics in 1603, nearly 40 years before Isaac Newton

RENÉ DESCARTES, 1596–1650

Descartes was born in La Haye en Touraine in France, the son of a local politician. His mother died when he was only one year old. Although he initially followed his father's wishes in training in law and science, Descartes abandoned the plan to become a lawyer and spent his time studying mathematics, philosophy, and science, in independent thinking, and in forays into the army. Luckily, he had sufficient wealth to support this lifestyle. He has been called the "father of modern philosophy," and his development of Cartesian coordinates was named by the British philosopher John Stuart Mill (1806–73) "the greatest single step ever made in the progress of the exact sciences." For the history of physics, Descartes' most important philosophical development was the mechanistic model—he endeavored to see the whole universe as machine-like systems that followed a system of physical laws in their operation.

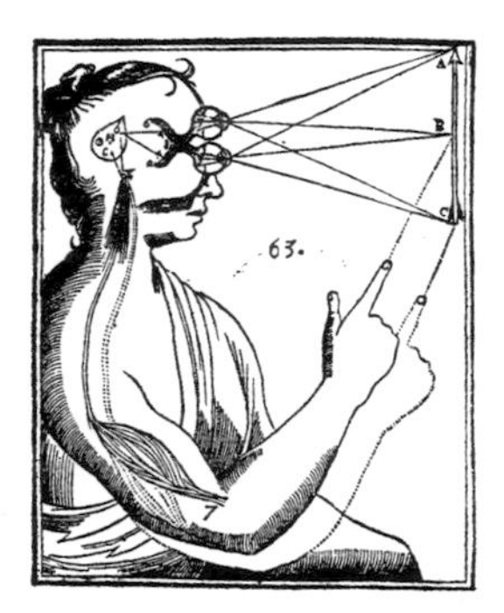

Descartes' model of vision showed how light rays travel to the eyes and information is transmitted to the pineal gland.

Descartes was sensitive and comfort-loving from childhood. He rose late, and said his best work was done in a comfortable bed (as was his development of the Cartesian coordinate system, see box, page 45). When the young Queen Christina of Sweden employed him as her tutor and insisted on tutorials at 5 AM in a freezing library, it took only five months for Descartes to fall prey to a serious lung condition and die, aged just 46.

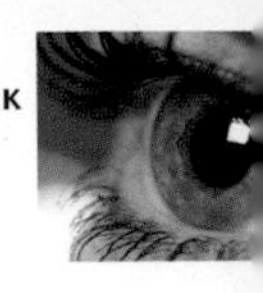

was born. Although the first astronomical telescope was made by Leonard Digges in England in the early 1550s (see page 159), they are most closely associated with the work of one man, the astronomer Galileo Galilei (see box, page 41) .

Through a Glass Brightly

Lenses change the path of light; they are the most basic optical tool. They were developed long before anyone could explain them. The earliest surviving example is the Nimrud lens, made in Ancient Assyria 3,000 years ago from a piece of rock crystal. Similar lenses were used in Babylon, Ancient Egypt, and Ancient Greece, perhaps to magnify objects or as burning lenses, focusing rays of sunlight to start a fire. While the Greeks and Romans filled spherical glass vessels with water to make lenses, glass lenses ground to the required shape were not developed until the Middle Ages.

The first use of a lens to correct vision may have been recorded by the Roman author Pliny the Elder (23–79 CE) who reported that Nero watched the gladiatorial games at the Colosseum through an emerald. Reading stones—convex lumps of glass or rock crystal—were used to magnify text from the 11th century. Ground glass lenses were used in spectacles from around 1280, though no one knew at first how or why they worked. The development of the microscope and the telescope during the 16th and 17th centuries generated a need for more accurate lenses. As grinding techniques were refined over the centuries, improved lenses led to further discoveries that then prompted demand for even better lenses. Some of the greatest scientists of the Renaissance and the Enlightenment, including Galileo, the Belgian microscope pioneer Antonie van Leeuwenhoek (1632–1723), and the Dutch physicist and astronomer Christiaan Huygens (1629–95), made their own lenses.

The Nimrud lens, discovered in Kurdistan (northern Iraq)

Pressure in the Aether

René Descarte's work on optics described the working of the eye and suggested improvements to the telescope. He used mechanical analogies to derive many properties of light mathematically, including the laws of reflection and refraction. In other regards, though, he was rather hampered by his refusal to accept the existence of a void. For theorists such as Gassendi, who envisaged a void with moving atoms, light could be explained as a stream of fast-moving particles that hurtled through space. With no void, Descartes needed a different mechanism. He believed that some type of thin "interstitial fluid"—another version of the aether—filled all the spaces, and that it was pressure exerted through this fluid that produced vision. So, if the sun pushed on the interstitial fluid, this pressure would be instantly transmitted to the eye, which would then perceive the sun. There was little basis for this theory, especially once we consider that the sun is separated from the Earth by 19.3 million miles (150 million km), but it did lay the

> *"[Descarte] kept a good conditioned handsome woman that he liked, and by whom he had some children (I think two or three). 'Tis a pity, but coming from the brain of such a father, they should be well cultivated. He was so eminently learned that all learned men made visits to him, and many of them would desire him to show them his store of instruments (in those days mathematical learning lay much in the knowledge of instruments, and as Sir Henry Savile said, in doing of tricks). He would draw out a little drawer under his table and show them a pair of compasses with one of the legs broken; and then for his ruler, he used a sheet of paper folded double."*
>
> John Aubrey, *Brief Lives*

foundation for a far more important piece of work by Christiaan Huygens (see page 48), the son of a close friend of René Descartes, and it prompted Newton to pursue his own ideas on the subject but in a different direction.

Lord of Light: Isaac Newton

Newton was possibly the greatest scientist who has ever lived; he was to become the giant on whose shoulders others have stood for more than 400 years. His work on forces and gravity (see page 74) is perhaps more famous than his work on optics, but not more important.

Newton successfully split white light into its constituent spectrum and then recombined the colored rays into white light, so demonstrating conclusively that white light is a mixture of colors. This possibility had been noted much earlier. Aristotle claimed that a rainbow is caused by clouds acting as a lens on sunlight, an explanation that was also accepted by al-Haytham. The Roman philosopher Lucius Annaeus Seneca (*ca.*55 BCE–*ca.*40 CE) referred in *Naturales quaestiones* to producing a band of colors similar to those of a rainbow by passing sunlight through glass prisms. In Newton's time, though, most people believed colored light was a form of shadow, made by mixing white light with darkness. Descartes thought color was caused by the spinning motion of the particles that made up light. Newton's

Through the prism of genius: Isaac Newton's work on gravity and optics revolutionized natural philosophy.

> *"Nature and Nature's laws lay hid in night; God said 'Let Newton be' and all was light."*
>
> Alexander Pope, 1727

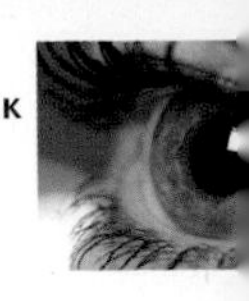

A FLY THAT MADE HISTORY

Descartes gave his name to the system of Cartesian coordinates still used to specify a point in three-dimensional space by relating its location to three axes—x, y, and z. He claimed to have developed the system in 1619 while lying in bed watching a fly buzzing in a corner of his bedroom. He realized that the fly's position could be identified precisely at any given moment by plotting its distance from the two nearest walls and either the floor or ceiling—in other words, its coordinates in three dimensions. From this simple observation it followed that a geometric shape could be represented by numbers (the coordinates of its corners) and that a curve could be described by a series of numbers relating to each other in an equation (hence a parabolic path can be plotted as a graph, for instance). The entire system of geometry investigated through algebra became possible once Descartes had seen and pondered his cornered fly.

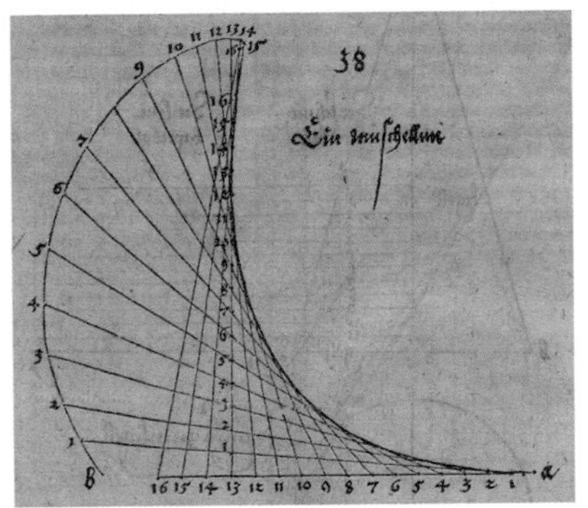

By plotting a series of points in terms of their distances from two axes, Cartesian geometry shows an equation as a graph.

When Newton's color wheel is rotated very rapidly, the colors are indistinguishable and the wheel looks white.

great intellectual rival Robert Hooke thought color was imprinted on light, as when it shines through a stained-glass window. He tried to use a prism to split light, but produced only white light with colored edges. Newton succeeded where Hooke failed because he used superior equipment. He made a pinhole in a black window screen to let a narrow beam of light into his room in Trinity College, at Cambridge University, and used an accurately crafted glass prism to split the beam, casting an image on to another screen several feet away. By allowing enough space for the colored beams to spread out properly, he produced a clear spectrum.

Newton took his dedication to experimental optics beyond the bounds of what was really sensible. In a famous

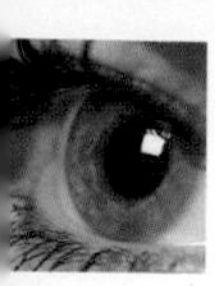

> *"If I have seen farther, it is by standing on the shoulders of Giants."*
>
> Isaac Newton, in a public letter to Robert Hooke, written at the insistence of the Royal Society to heal —or cover up—the rift between the pair

account of self-harm, he poked a bodkin (a large, blunt needle) into his eye socket, pressing it as far back as he could without puncturing the eyeball itself, in an attempt to distort the shape of the eyeball and see how this affected his perception of color. Newton realized that colored objects appear the color they do because of the light they reflect. For example, a red cape appears red because it reflects red light, whereas a white shirt reflects all light. He also associated different degrees of refrangibility with different colors.

Despite this admirable dedication to science, Newton was a difficult, arrogant, and argumentative man. His antagonism toward Hooke was obsessive, but not unique; several other people aroused his hatred and vitriol, too. Hooke's fame would have been greater if he had not had the misfortune to die before Newton, who appropriated one of Hooke's discoveries, the so-called Newton's rings of color seen in thin films of oil on water. Indeed, Newton deliberately withheld his own work on light and color, *Opticks*, from publication for 30 years, only releasing it after Hooke was dead and so unable to dispute authorship.

> *"I tooke a bodkine gh & put it betwixt my eye & [the] bone as neare to [the] backside of my eye as I could: & pressing my eye [with the] end of it (soe as to make [the] curvature abcdef in my eye) there appeared severall white darke & coloured circles r, s, t, &c. Which circles were plainest when I continued to rub my eye [with the] point of [the] bodkine, but if I held my eye & [the] bodkin still, though I continued to presse my eye [with] it yet [the] circles would grow faint & often disappeare untill I removed [them] by moving my eye or [the] bodkin.*
>
> *If [the] experiment were done in a light roome so [that] though my eyes were shut some light would get through their lidds There appeared a greate broade blewish darke circle outmost (as ts), & [within] that another light spot srs whose colour was much like [that] in [the] rest of [the] eye as at k. Within [which] spot appeared still another blew spot r espetially if I pressed my eye hard & [with] a small pointed bodkin. & outmost at vt appeared a verge of light."*
>
> Newton's notebook, CUL MS Add. 3995

HOOKE'S MICROGRAPHIA

Hooke's most famous work is the *Micrographia*, published in 1665. This was a fine example of how developments in optics led quickly to developments in other areas of science, principally biology and astronomy. Although Hooke was not the first microscopist, he brought microscopy into mainstream science and made improvements to both microscope and telescope design. *Micrographia* features drawings of objects, organic

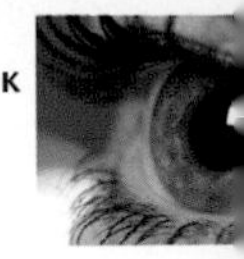

materials, and tiny organisms seen through Hooke's microscope. The detailed illustrations (some drawn by the architect Christopher Wren) were ground-breaking, making *Micrographia* one of the most important science books ever published. Samuel Pepys records in his diary that he sat up until 2 AM reading it, and that it was "the most ingenious book that ever I read in my life."

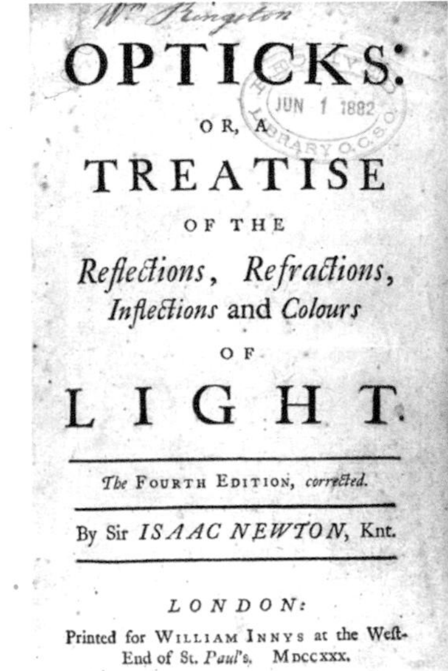

OPTICKS:
OR, A
TREATISE
OF THE
Reflections, *Refractions*,
Inflections and *Colours*
OF
LIGHT.

The FOURTH EDITION, *corrected.*

By Sir *ISAAC NEWTON*, Knt.

LONDON:
Printed for WILLIAM INNYS at the West-End of St. *Paul's*. MDCCXXX.

Title page of Newton's treatise on optics, published in 1704

WAVE OR PARTICLE?

It's one thing to recognize that white light is a composite of colored light, but this then raises the question of what colored light is. Differing opinions about whether light is made up of particles or is a wave of some type are found in early Indian writings on science. In Europe, Empedocles suggested rays and Lucretius spoke of particles, and the debate continued through the centuries; Hooke, following Descartes, took the view that light is a form of wave. This was yet another point of dispute with Newton, who wrote of "corpuscles" (i.e. particles) of light, an idea first proposed by Gassendi and that Newton read about in the 1660s. Newton was so influential that wave theory found little favor in Britain for a long time. Elsewhere in Europe, however, Newton's arrogance and argumentative nature made him unpopular and if anything damaged support for the corpuscular model. Newton rejected wave theory because he believed that a longitudinal wave (vibrating in the direction of propagation) could not account for polarization. No one had considered the possibility of transverse waves (which vibrate in a direction at right angles to the direction of propagation). Newton accepted the idea of a luminiferous (light-bearing) aether, a medium through which light traveled, although this was not strictly necessary for his corpuscular theory as the particles could have traveled equally well through a void. He also believed that the corpuscles of light switched between two phases known as "easy reflection" and "easy transmission." Periodicity is a basic feature of wave theory,

Newton probed his own eye with a bodkin; an illustration from Newton's Opticks.

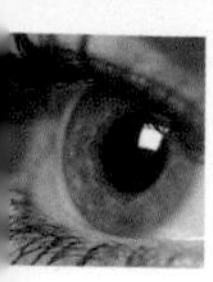

and in this he anticipated quantum mechanics (see page 116). While Newton's name is closely associated with corpuscular theory, his own writings incorporate aspects of both ideas. For example, he explained diffraction by suggesting that light corpuscles created localized waves in the aether. Interestingly, this places him closer to the modern view of the "duality of light"—that it has qualities of both a wave and a particle.

Depiction of a magnified flea from Hooke's Micrographia

> *"[Hooke] is but of middling stature, something crooked, pale faced, and his face but little below, but his head is large; his eye full and popping, and not quick; a grey eye. He has a delicate head of hair, brown, and of an excellent moist curl. He is and ever was very temperate, and moderate in diet etc.*
>
> *As he is of prodigious inventive head, so is a person of great virtue and goodness. Now when I have said his inventive faculty is so great, you cannot imagine his memory to be excellent, for they are like two buckets, as one goes up, the other goes down. He is certainly the greatest expert on mechanics this day in the world. His head lies much more to geometry than to arithmetic. He is a bachelor, and, I believe, will never marry. His elder brother left one fair daughter, which is his heir. In fine, (which crowns all) he is a person of great suavity and goodness.*
>
> *It was Mr Robert Hooke that invented the pendulum watches, so much more useful than the other watches.*
>
> *He has invented an engine for the speedy working out of division, etc or the speedy and immediate finding out of the divisor."*
>
> John Aubrey, *Brief Lives*

Wave-Fronts and Quanta

In Europe, Christiaan Huygens developed the wave-front theory. His theory of light, which he completed in 1678 but didn't publish until 1690, was based on his own experimental findings. Like Descartes, who had been a regular visitor to Huygens' childhood home, he considered light to be a wave propagated through the aether. He predicted that light would travel more slowly through a dense medium than a less dense medium. This was significant in that—unlike Descartes—he was saying that the speed of light is finite.

Huygens' wave-front theory explains how waves evolve and behave when they encounter obstacles—being reflected, refracted, or defracted. He suggested that each position in a wave becomes the center of a wavelet traveling out in all directions. In the case of light, which he considered a pulse phenomenon, repeated waves were emitted and traveled outward at the speed of light. The light wave is propagated through three-dimensional space, in the form of a spherical wave.

At the edge of a region reached by the rays of light, the wavelets interfere with one another and may cancel each other out. If

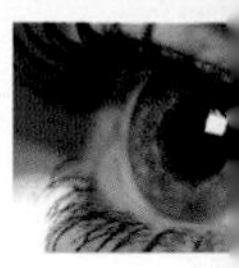

ROBERT HOOKE (1635–1703)

Robert Hooke was born on the Isle of Wight, England, where his father was a church curate. Hooke went to Westminster School in London at the age of 13 when his father died, and then to Christ Church College, Oxford, as a chorister. Had he been healthier, Hooke was destined for a career in the Church, but turned to science instead, becoming assistant to the chemist Robert Boyle in Oxford. Hooke moved back to London in 1660 and became a founder member of the Royal Society in 1662. As the Society's first curator, Hooke was charged with demonstrating "three or four considerable experiments" each week. He made extensive studies with his microscope, publishing drawings of what he saw in the *Micrographia* (1665) and coining the term "cell" for the components of living tissue (so named because the "pores" he saw in a sliver of cork reminded him of the rooms or "cells" that monks occupied).

Hooke was one of the two Surveyors of London employed after the city was destroyed in the Great Fire of 1666, a post that made him rich. He also built Bethlem Royal Hospital—the infamous London lunatic asylum better known today as "Bedlam."

An ingenious thinker, experimental scientist, and mechanic, he came up with innovations and improvements to many existing devices, including the air pump, microscope, telescope, and barometer, and he pioneered the use of springs to power clocks. Most of his ideas were developed further by other people, Hooke having provided the essential springboard but getting little of the credit. He had theories on combustion and gravity, even suggesting the inverse square law in relation to gravity in 1679—the cornerstone of Newton's own work on the subject. Newton never allowed any suggestion of Hooke's precedence or brilliance, and the shadow of Newton's animosity denied Hooke the place in history he deserved. There is no known surviving portrait of Hooke.

Part of the devastated landscape of London after the Great Fire of 1666

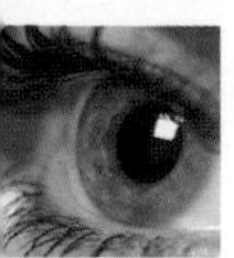

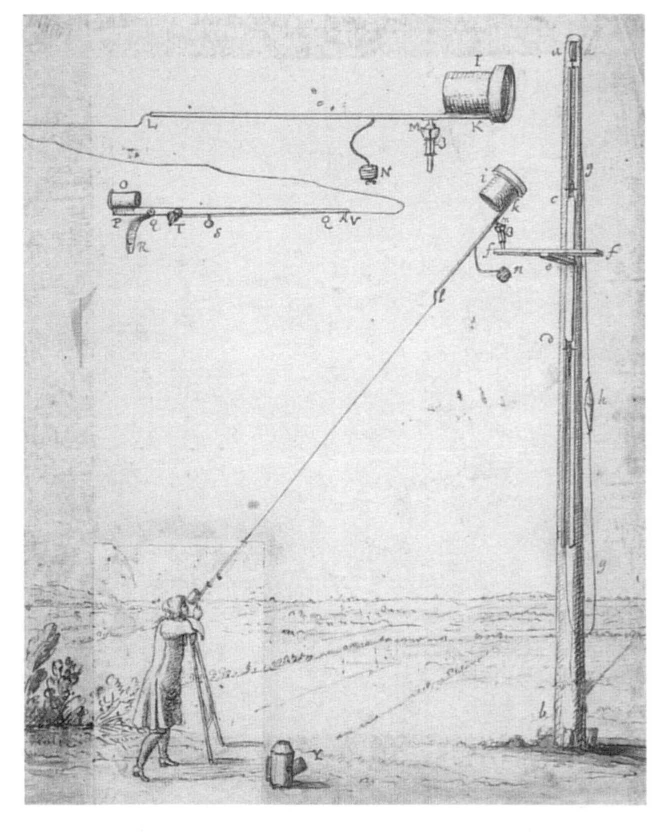

Huygens' aerial telescope has a long focal length, achieved by separating the objective and the eyepiece, then using a string to align them.

they hit an opaque object, parts of the wavelet are cut off and some persist, producing the complex fine structure of lines at the edges of shadows and images that form diffraction patterns. Scientific opinion is divided as to whether Huygens discovered the principle through a stroke of genius or just good luck and that he got the right answer for the wrong reasons.

During the 19th century, several scientists working in different European countries established the theory that light is a transverse wave (vibrating at a right angle to the direction of propagation and travel, like a snake wriggling across the ground). In 1817, the French physicist Augustin-Jean Fresnel (1788–1827) presented his own wave theory of light to the Académie des Sciences and by 1821 he had shown that polarization could be explained only if light comprised transverse waves, with no longitudinal vibration. This answered Newton's principle objection to regarding light as a wave. Fresnel is best known as the inventor of the lens that bears his name, originally designed to enhance the beam shining from lighthouses.

Christiaan Huygens, 1671

Thomas Young

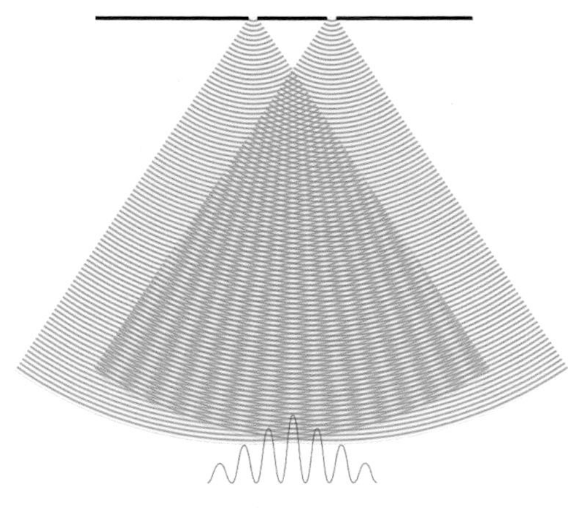

The interference pattern produced when light shines through two slits, supported the wave theory of light.

Young's Double-Slit Experiment

In 1801, Thomas Young conducted an experiment that seemed to prove once and for all that light is a wave. He shone light through two slits he had made. Rather than seeing the sum of the results from experiments with single slits, as expected, he noticed a complex diffraction pattern, caused by interference between the light from the two slits. The more slits he added, the more complex the interference pattern became. This demonstrated that light is indeed a wave, with the troughs and peaks of the waves either canceling each other out or reinforcing each other to make interference patterns. Young also proposed that different colors of light are the result of differing wavelengths, a small step toward the realization that would come later in the 19th century—that the light we see is just one part of a whole spectrum of electromagnetic radiation (EMR), now known to comprise gamma-rays, X-rays, ultraviolet light, visible light, infrared, microwaves, radio waves, and long waves.

A New Dawn—Electromagnetic Radiation

It was James Clerk Maxwell (1831–79) who first showed that electromagnetic radiation consists of transverse waves of energy moving at the speed of light. The different types of electromagnetic radiation—including light and radio waves—are characterized by different wavelengths. In fact, the English physicist Michael Faraday (1791–1867) had already demonstrated the connection between electromagnetism and light in 1845 when he showed that the plane of polarization of a beam of light is rotated by a magnetic field (see box, page 105).

Maxwell still assumed there was a luminiferous aether through which all

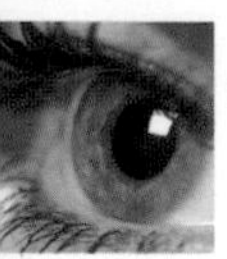

forms of electromagnetic radiation must move. The aether was unlike anything else in that it was a true continuum—it was infinitely divisible and not made of discrete particles like normal matter. Not only was the aether infinitely divisible, so were the waves of energy that traveled through it. There were problems with Maxwell's theory that were only resolved when Max Planck showed that energy must be emitted in tiny but finite amounts, now called quanta. (Otherwise, for complex reasons, all the energy in the universe would be transformed into waves of high frequency.)

Albert Einstein demonstrated in 1905 in his work on the photoelectric effect (see box, opposite) that light itself behaves as though it is made up of quanta, or tiny packets of energy, now called photons. He used what we now refer to as Planck's constant to relate the energy of a photon to its frequency.

James Clerk Maxwell

Light is now considered to have wave-particle duality: it sometimes behaves like a wave and sometimes like a particle. It is useful to have some way of predicting when it will do one or the other and quantum mechanics can predict just this (see page 117).

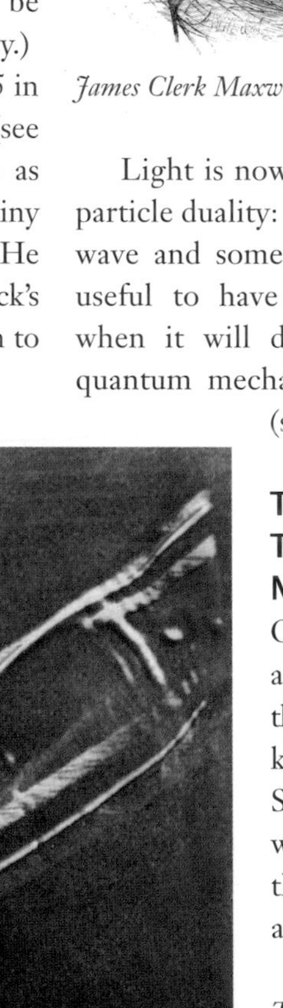

The End of an Aether: The Michelson–Morley Experiment

Our normal understanding of a wave is that it has to travel through a medium of some kind, such as air or water. Similarly, it was assumed that waves of light must travel through the luminiferous aether in a similar way.

The first color photograph ever produced was taken by James Clerk Maxwell in 1861 and shows a tartan ribbon.

THE PHOTOELECTRIC EFFECT

An early photoelectric cell was produced in the development of television.

When Albert Einstein was awarded the Nobel Prize for Physics in 1921, it was not for his most famous ideas—the theories of relativity—but for his work on the photoelectric effect. He explained how a photon (though not called that at the time) could sometimes knock an electron out of its orbit in an atom, generating a tiny burst of energy. This is how photoelectric solar power panels generate electricity from sunlight. The electrons that sunlight knocks out of a piece of semiconductor material such as silicon can be directed to flow along a wire and then be siphoned off to do useful work or be stored for later use. The photoelectric effect was first recorded by the French physicist Alexandre Becquerel (1820–91) in 1839. He observed that when blue or ultraviolet light is shone on to certain metals it generates an electric current, but he did not know how it worked. Einstein borrowed Max Planck's idea of quanta, originally applied to the energy of atoms, and used it to describe little packets of light energy—photons. The amount of energy a photon represents depends on the wavelength of the light. Whereas photons of blue light have enough energy to knock an electron out of its orbit and thereby free it, generating an electric current in the process, photons of red light do not. Increasing the intensity of the red light doesn't help as the individual red-light photons are just not up to the job.

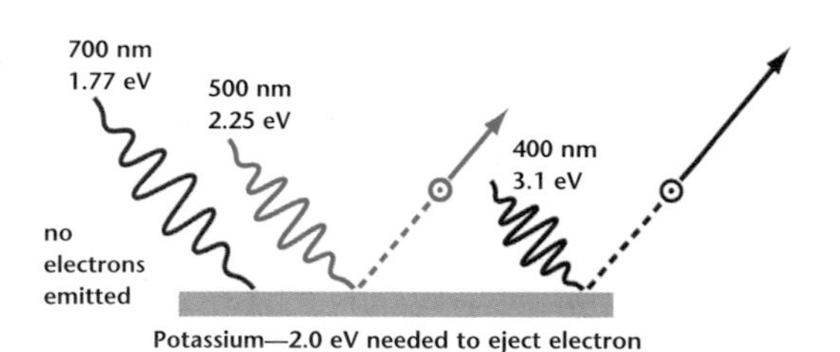

Photons falling on the surface will only knock out an electron if they have sufficient energy; red light will not produce a current, but blue or green light will.

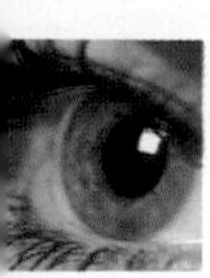

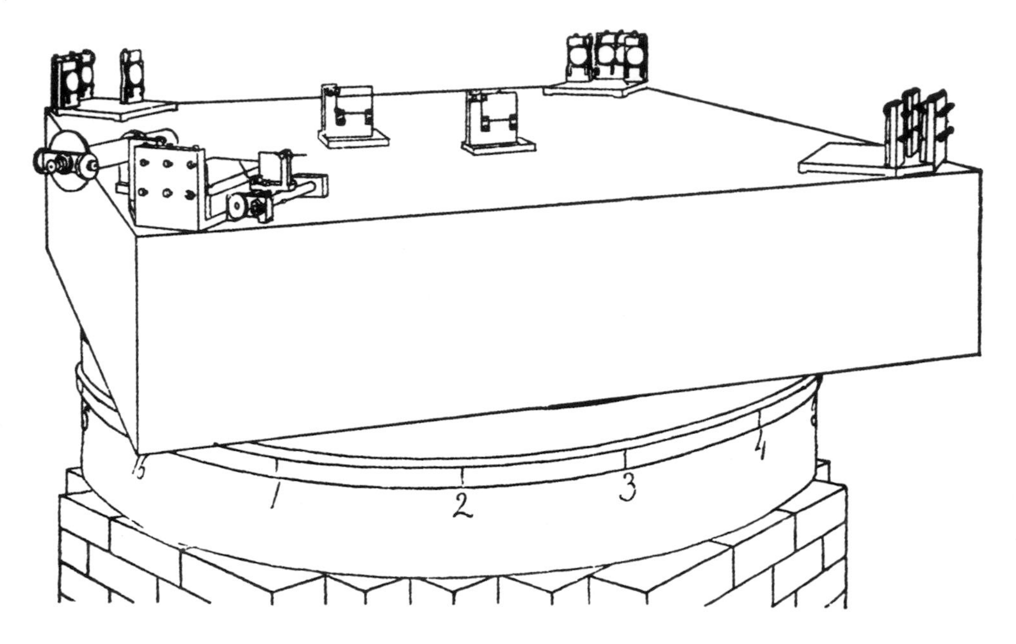

Michelson–Morley's equipment for measuring the speed of light was designed with the intention of proving the existence of the aether.

The end for aether eventually came as a result of an experiment carried out in 1887 by two American physicists, Albert Michelson (1852–1931) and Edward Morley (1838–1923). If the aether existed, scientists assumed, it must fill space as it carried light from the sun and stars to the Earth. In 1845, the British physicist George Gabriel Stokes (1819–1903) had suggested that as the Earth is moving at great speed through space there should be an effect due to drag from our planet as it passes through the aether. At any point on the Earth's surface, the speed and direction of the aether "wind" would vary depending on the time of day and year, so it should be possible to detect the motion of the Earth relative to the aether by looking at the speed of light at different times and in different directions.

Michelson and Morley built equipment to measure the speed of light so precisely that it would detect the effect of the aether, if present. Their apparatus split a beam of light into two beams that traveled at right angles to each other toward two mirrors. The beams were reflected backward and forward for a distance of 36 feet (11 meters) before being recombined in an eyepiece. If the Earth were traveling through aether, a beam moving parallel to the flow of aether

> *"[The aether] is the only substance we are confident of in dynamics. One thing we are sure of, and that is the reality and substantiality of the luminiferous ether"*
>
> William Thomson, Lord Kelvin, 1884

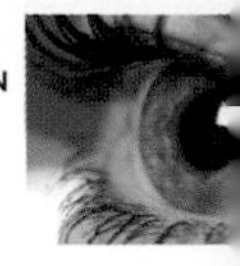

Early scientists could not agree whether the Earth hangs in empty space or moves through an aether.

The Michelson interferometer (see page 56) can be used to produce colorful interference from white light.

would take longer to return to the detector than a beam moving perpendicular to the aether. If one beam traveled more slowly than the other, this should show up in the interference fringes produced when the beams recombined. The whole apparatus was built on a block of marble, floating in a bath of mercury, set in the basement of a building to eliminate as far as possible any vibration that might interfere with the results. The equipment was more than sensitive enough to detect the effect that would be expected if the Earth really were subject to wind from the aether. When it failed to show statistically relevant positive results, Michelson and Morley had to report

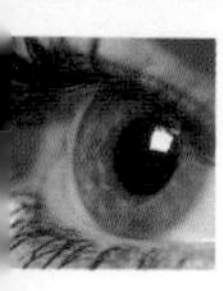

A Michelson interferometer works by splitting a beam of light in two, then reflecting and recombining the resulting beams.

that their experiment was a failure. Others went on to refine the apparatus, but still did not find evidence of aether. The Michelson and Morley experiment had not, of course, failed. It had shown that there is no luminiferous aether. Unfortunately, Michelson's conclusion was not that there is no aether, but that the model of a stationary aether that imparts drag on light (the aether-drag hypothesis), proposed by Augustin-Jean Fresnel, was the correct one.

Jupiter and its moon Io; eclipses of Jupiter by its moons convinced Huygens that light travels at a finite speed.

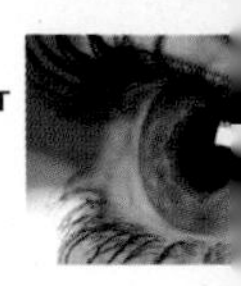

C

The speed of light is represented by the letter "c" (as in $E=mc^2$), standing for the Latin *celeritas,* meaning swiftness or speed.

At the Speed of Light

As early as *ca.*429 BCE, Empedocles believed that light travels at a finite speed, though it seems to arrive instantaneously. He was, though, a notable exception among the ancient thinkers, for most agreed with Aristotle that the speed of light was infinite. The Arab scientists Avicenna and al-Haytham agreed with Empedocles, as did both Roger and Francis Bacon. But the prevailing view even in 17th-century Europe, and that held by Descartes, was that light travels at infinite speed.

The first attempt to challenge this assumption and measure the speed of light was made by Galileo in 1667 using a very primitive method. Galileo and an assistant standing 1 mile (1.6 kilometers) away took turns covering and uncovering lanterns and timing how long it took for them to notice the light. It was probably a better measure of the speed of their reactions than anything else. Galileo concluded that if the speed of light was not infinite it was certainly very fast—probably at least 10 times the speed of sound, which had first been measured by the French philosopher and mathematician Marin Mersenne (1588–1648) in 1636.

Huygen's conviction that light travels at a finite speed followed observations made by the Danish scientist Ole Rømer (1644–1710), working with the Italian-born astronomer Giovanni Cassini (1625–1712) in Paris, after viewing eclipses of the moons of Jupiter. Cassini and Rømer noticed that although the eclipses should occur at regular intervals they were not always on time—and the variation depended on the position of the Earth relative to Jupiter. They concluded that when Earth is further from Jupiter, we see the eclipse later because it takes longer for the light to reach Earth. Cassini stated in 1676 that discrepancies in the apparent times of these eclipses could be explained if light traveled at a finite speed. He went on to calculate that it would take about 10 or 11 minutes for light to travel from the sun to the Earth. He didn't pursue the issue further though, and it was left for Rømer to calculate the speed of light precisely. He correctly predicted the exact time of an eclipse of Io in 1679, at 10 minutes later than everyone expected. Working from the best estimate of the diameter of the Earth's orbit, he calculated the speed of light as 124,274 miles (200,000 km) per second. Using the current figure for

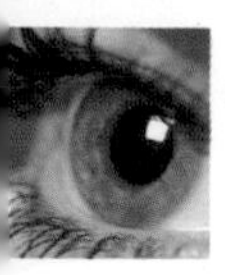

the Earth's orbit in Rømer's formula gives 185,169 miles (298,000 km) per second, very close to the modern value of 299,792.458 km per second. (This speed won't be changed by future work, as the length of a meter is set as the distance light travels in 1/299,792,458 of a second.)

In 1678, Huygens used Rømer's method to show that light takes a matter of seconds to travel from the moon to Earth. Newton

ARCHIMEDES' HEAT RAY

According to tradition, the Greek scientist, mathematician, and engineer Archimedes (*ca.*287–*ca.*212 BCE) set up a parabolic arrangement of mirrors on the shore to use sunlight to set fire to enemy ships during the Siege of Syracuse (*ca.*214–212 BCE). An experiment in 1973 at a naval base near Athens used 70 copper-coated mirrors 5 ft (1.5 m) by 3.3 ft (1 m) across to direct sunlight at a plywood mock-up of a Roman warship painted with tar some 164 ft (50 m) distant. The ship burst into flames within seconds. A similar experiment in 2005 by a group of students from Massachusetts Institute of Technology (MIT) also set fire to a mock ship under perfect weather conditions.

Although this technique, like using a convex lens to start a fire, apparently uses light, it is not, of course, visible white light that sets fire to ships or kindling but the invisible infrared radiation (heat) that accompanies it in sunlight.

This illustration shows Archimedes burning an enemy ship with the use of a mirror; in fact it took more than one mirror!

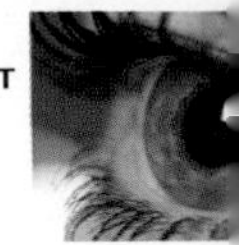

stated in the *Principia* that light takes seven or eight minutes to reach Earth from the sun, which is quite close to the actual figure of 8 mins 20 secs, on average.

Newton and others assumed that the speed of light varied depending on the medium through which it traveled. If light were particles, this would make sense. If light is a wave, it is not necessarily the case. Not everyone was convinced by Huygens' calculations, and opinion over whether light traveled at a finite speed remained divided until the English astronomer James Bradley (1693–1762) decided the question once and for all in 1729. He discovered the aberration of light (also called stellar aberration). This is the phenomenon of a star appearing to describe a small circle about its true position as a result of the Earth's velocity (speed and direction) in relation to the star. His study took more than 18 years to complete.

Two later French experimenters recreated Galileo's experiment with lamps and servant but in a rather more sophisticated way. In 1849, the physicist Hippolyte Fizeau (1819–96) used two lanterns and a rapidly rotating wheel with teeth that alternately obscured and revealed the light, and a mirror to reflect it back. The light could only shine back through the same gap if it returned quickly enough, so its speed could be calculated from the speed of the wheel's rotation. By rotating a wheel with a hundred teeth at several hundred times a second he was able to measure the speed of light to within about 994 miles (1,600 km) per second. Léon Foucault (1819–68) used a similar principle. He shone a beam of light on to a rotating and tilting mirror, then bounced it back from a second mirror placed

CLOAK OF INVISIBILITY

During the 1990s, scientists developed meta-materials with a negative refractive index. The refractive index of a material determines how much of the light entering it will be refracted. A vacuum has a refractive index of one, and denser materials have higher refractive indices. In 2006, meta-materials were used in the first cloaking device, making an object appear invisible to microwaves. The particles of meta-material must be smaller than the wavelength of light so that light will flow around them, like water flowing around a rock in a stream. So far, a cloaking device that works with light waves and is more than a few microns across has not been perfected.

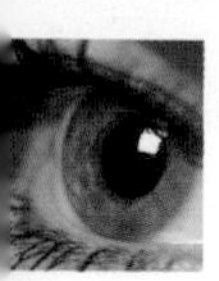

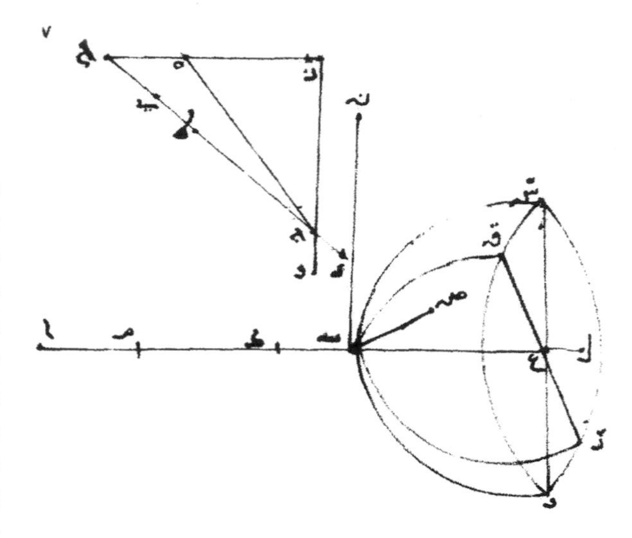

The original manuscript featuring Ibn Sahl's explanation of the law of refraction, ca.984.

22 miles (35 km) away. As the angle of the rotating mirror changed, he was able to calculate the angle at which the returned light was re-reflected and therefore determine how far the mirror had moved and how much time had passed. In 1864 Fizeau suggested that the "length of a light wave be used as a length standard," and in redefining the meter in terms of the speed of light that has effectively been achieved.

Einstein based his theories of relativity on the observation that the speed of light is constant throughout the universe.

Straight and True

Anaxagoras was already certain in the fifth century BCE that light travels in straight lines only, and this belief held until the 20th century when Einstein said that light could be bent into a curved path by gravity. It was clear to the ancients, though, that light can be made to change direction—when it is reflected, for instance, or when it is refracted as it moves from one medium into another. Ptolemy gave an approximate account of refraction, and it was described in 984 by the Persian physicist Ibn Sahl (*ca.*940–1000 CE).

However, the mathematical law explaining and predicting the angle of refraction is known as Snell's Law after the Dutch astronomer Willebrord Snellius (1580–1626). Although Snellius rediscovered the relationship in 1621, he didn't publish it. Descartes published a proof of the law in 1637. Snell's law works because, as the French mathematician Pierre de Fermat (1601–65) showed, light takes the quickest path through any substance.

That light follows a curved path was first confirmed in the early 20th century as part of a demonstration of Einstein's theory of relativity.

The astronomer Arthur Eddington led a British expedition to the Island of Principe off the African Coast to take advantage of a total solar eclipse visible there in 1919. The

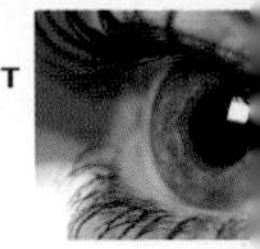

expedition photographed stars lying close to the sun's position that would otherwise be obscured by the sun's light. One star that actually lay behind the sun and so should have been hidden from view was clearly visible in one of the photographs that Eddington took. This demonstrated that the star's light had been bent by the gravitational field of the sun and this had altered the star's apparent position to a point where it was now visible.

Artist David Hockney painted a series of swimming pool paintings in which he played with the refraction and reflection of light hitting air and water.

Light in the EMR Spectrum

Light has occupied a special place in the story of physics because it is visible and it makes a world of difference to humankind. But as Maxwell's work demonstrated, visible light is only one form of electromagnetic radiation. All forms move at the speed of light, all are quantized forms of energy (that is, they can exist as either particles or waves), but visible light is the only one we see. There was no early attempt to distinguish the heat from the sun (its infrared radiation) from its visible light. Other forms of electromagnetic radiation such as X-rays, radio waves, and microwaves were not discovered until the late 19th century.

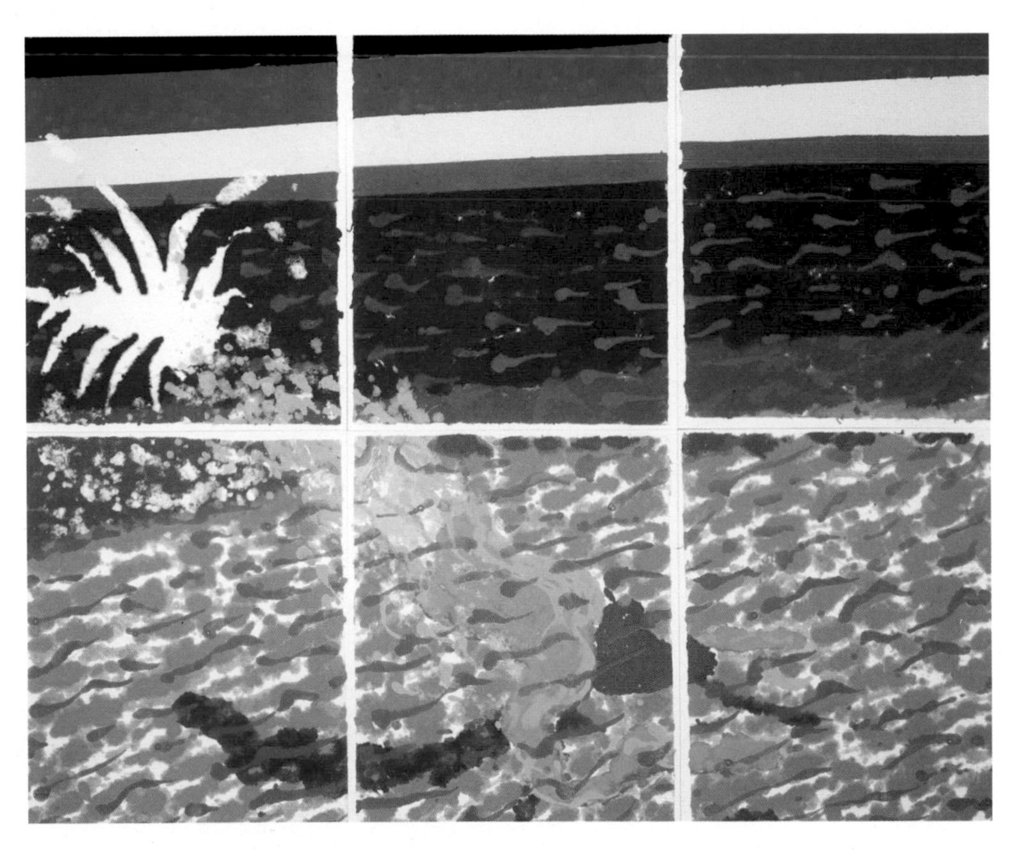

CHAPTER 3

Mass in Motion— MECHANICS

Mechanics is the term used to describe the way bodies act when subjected to forces. Classical mechanics began in earnest when Newton described his three laws of motion. It deals with the actions of bodies and matter of all types and sizes above the atomic, from ball bearings to galaxies, including liquids, gases, and solids, inanimate objects, and parts of living organisms. People were making practical use of physical forces long before they had any understanding of them, or even began to wonder about the laws that govern them. The earliest builders used levers and rollers to move large blocks of stone; they employed gravity to drop things into place and to check the perpendicular with plumb lines.

The harnessing of mechanical energy has enabled the development of the modern world.

The Ancient Egyptians may have used mechanical devices such as levers and rollers to help move the blocks of stone needed to build the pyramids.

Mechanics in Action

Whenever we make use of the forces that act on matter we are employing mechanical laws to work for us. The Egyptian builders of the pyramids didn't (as far as we know) have any understanding of the forces involved in moving blocks of stone to build the pyramids, nor did the architects of the complex irrigation systems used in Sri Lanka have formal knowledge of fluid dynamics. Yet both cultures were able, through experimentation, trial, and error, to use the laws of physics to their advantage.

The fertile crescent is an area that stretches from the Mediterranean to the Persian Gulf. It takes in all the land that falls between the rivers Tigris and Euphrates—known to the Greeks as Mesopotamia ("between two rivers")—including areas that are now Syria and Iraq. Farming developed in this area around 10,000 years ago, and by 5000 BCE the Sumerians had built the first cities, employing methods of cutting, moving, and stacking vast stone blocks. The Sumerians also invented the wheel, thus harnessing physical forces in a new way. As the populations grew, the

EARLY ENGINEERING AND HYDRAULICS IN ACTION

Hydraulic engineers in Sri Lanka built complex irrigation systems in the third century BCE. The system was founded on the invention of the *biso-kotuwa*, similar to a modern valve-pit, which regulates the outward flow of water. Vast dammed rainwater reservoirs, channels, and sluices provided enough water to sustain the Sinhalese people of Sri Lanka on a rice-based diet. The first rainwater tank was built in the reign of King Abhaya (474–453 BCE). Much more sophisticated and extensive systems were built centuries later, starting during the reign of King Vasaba (65–108 CE). His engineers built 12 irrigation channels and 11 tanks, the largest being 2 miles (3 km) across. Their greatest achievements were made under the rule of King Parakrambahu the Great (1164–96 CE), when Sinhalese engineers achieved a steady gradient of 32 inches per mile (20 centimeters per km) along irrigation channels that extended around 50 miles (80 km).

people of Mesopotamia made the first practical use of fluid dynamics, developing irrigation systems to water their farmland in the sixth millennium BCE.

Flowing water can be used for more than nourishing crops. It has its own force, and the pressure it exerts can be used to do useful work. The first known use of water to provide motive force was in Ancient China when Zhang Heng (78–139 CE) used water power to turn an armillary sphere (a globe used in astronomy to determine the positions of stars). Du Shi, in 31 CE, used a waterwheel to power the bellows in a blast furnace producing cast iron.

Ancient Greek Mechanics

Although early civilizations made practical use of mechanics, we have no record of systematic thought about or analysis of forces. The first evidence of abstract thinking about how and why forces act on objects comes from Ancient Greece. In the *Mechanica*, Aristotle investigated how levers make it possible to move great weights using little force. His answer was: "Moved by the same force, that part of the radius of a circle which is farthest from the center moves more quickly than the smaller radius which is close to the center."

Aristotle recognized this soon after the invention of a form of balance that had arms of unequal length. In an equal-arms balance, weights on one side must be balanced by equal weights on the other side. But with an unequal-arms balance, the weights can also be balanced by moving the fulcrum (the point at which the crossbar pivots) and by moving one weight along its arm. Theorizing about mechanical forces, then, came about only after a practical device had been devised that put these forces to use. The existence of the unequal-arms balance gave Aristotle the opportunity for observation and investigation.

Aristotle's discovery is the precursor of

The Great Ziggurat of Ur (now in Iraq) was built around 4,000 years ago and represents a considerable feat of engineering.

"Give me a place to stand on, and I will move the Earth."

Archimedes

the law of the lever for which Archimedes (*ca.*287–212 BCE) provided a proof around a century later (although the law was probably already well known before Archimedes confirmed it).

In its modern form, the proof states that weight-times-distance on one side of the fulcrum is the same as weight-times-distance on the other side:

$$WD = wd$$

Archimedes expressed this in terms of ratios, as he would not have accepted the multiplication of dissimilar measures (weight and distance). As ratios, the law of the lever takes the form:

$$W:d = w:D$$

The Archimedes screw is used to move water even in some present-day irrigation systems.

ARCHIMEDES' INVENTIONS

Archimedes put his knowledge of mechanics to good, practical use. King Hieron II commissioned him to design a huge ship, the first luxury liner in history, capable of carrying 600 people and with facilities that included garden decorations, a gymnasium, and a temple dedicated to Aphrodite. To pump out any water that leaked into the hull, he is said to have developed the Archimedes screw, a revolving screw-shaped blade that fits snugly inside a cylinder and is turned by hand. The same design was adapted to transfer water from a low-lying source into irrigation canals and is still used today. Other inventions attributed to Archimedes include a parabolic arrangement of mirrors to focus the rays of the sun on enemy ships and burn them (see page 58), and a giant claw for lifting enemy ships out of the water. As so often, war seems to have provided the impetus for scientific developments.

Archimedes reputedly boasted that if he had a good enough lever and a place to stand, he could move the Earth. In principle, it's true.

The Problem of Dynamics

An arrow shot upward follows a predictable parabolic path.

Aristotle began with the proposition that something moves because a force is applied to it, and it keeps moving for as long as the force continues. The tendency of a moving body to keep going is now called momentum. This proposition of Artistotle's explains what happens if we push or pull something along, but clearly falls down when applied to projectiles. If we throw something, fire an arrow from a bow, or a bullet from a gun, the object continues moving after the thing or person initiating the "push" no longer has contact with the projectile. Aristotle solved the problem by transferring the status of "mover" to the medium through which the projectile travels, so the air continues to exert a force on the arrow, pushing it along toward its target. This force is impressed upon the air when the arrow is first released from the bow.

The Greek mathematician Hipparchus (*ca.*190–*ca.*120 BCE) rejected this, maintaining that the force had been transferred to the projectile itself. So, an arrow shot straight upward has more power—or impetus—to carry it away from earth than gravity has to pull it back to earth. This power naturally decays with time, however. It decays of itself, not because of air resistance, gravity, or any other influence. At the point where the impetus is equal to the pull of gravity, the arrow is momentarily still. It then starts to fall, the speed of its fall increasing as its original impetus decays to zero. As the impetus decays, it can do less to resist the pull of gravity on the object. When there is no residual impetus left, the arrow falls at the same speed as an object that has been dropped rather than thrown. Hipparchus' model also explained the behavior of a falling or dropped body. The object begins in a state of equilibrium between the downward pull of gravity and the upward thrust of the hand. The upward thrust is refreshed at the moment the object is released, but then steadily decays, so the object accelerates toward the ground. The model accounts for terminal velocity, too, since the rate of fall becomes steady once all the impetus of the body has decayed.

The philosopher John Philoponus (490–570 CE), sometimes known as John the Grammarian or John of Alexandria, had a similar theory of impetus. He suggested that a projectile has a force bestowed upon it by the "mover" but this is self-limiting and after it is expended the projectile returns to the pattern of normal movement. In the 11th century, Avicenna (*ca.*980–1037) found fault with Philoponus' model, saying instead that a projectile is given an inclination rather than a force and that this does not naturally decay. In a vacuum, for

STATIC MECHANICS

While the Ancient Greeks concerned themselves with dynamics (the mechanics of movement), the Romans mastered static mechanics. Static mechanics explains how forces held in equilibrium will hold a mass stationary. This is a fundamental principle in architecture, where unbalanced forces can cause a building or bridge to collapse. An arched bridge, for example, stays up simply because the pressure exerted by the stones making up the arch are perfectly balanced. The challenges of medieval and Renaissance architecture to build great vaulted ceilings, arches, and domes were problems in static mechanics that led to exquisite solutions.

The Duomo of the cathedral in Florence, built by Filippo Brunelleschi, represents a triumph of engineering—it is held up only by the weight of its own stone.

example, the projectile would move forever following the inclination bestowed upon it. In air, the resistance of the air eventually overcomes the inclination. He believed, too, that a projectile is pushed along by the movement of the air it displaces.

The Spanish-Arab philosopher Averroes (1126–1198 CE) was the first person to define force as "the rate at which work is done in changing the kinetic condition of a material body" and to argue "that the effect and measure of force is change in the kinetic condition of a materially resistant mass." He introduced the idea that non-moving bodies have resistance to starting to move—known today as inertia—but he applied it to the celestial bodies only. It was Thomas Aquinas who extended the concept to earthly bodies. Kepler followed the Averroes-Aquinas model—it was he who introduced the term "inertia"—which eventually became the central concept of Newton's dynamics. This means Averroes is responsible for one of the two crucial innovations in the development of Newtonian dynamics from Aristotelian dynamics.

The French philosopher Jean Buridan (*ca.*1300–*ca.*1358) related the impetus imparted by the mover to the velocity of the moving body. He thought that the impetus could be in either a straight line or a circle, the latter explaining the movements of the planets. His account is similar to the modern concept of momentum.

Buridan's pupil, Albert of Saxony (*ca.*1316–1390), expanded on the theory,

dividing the path of a projectile into three stages. In the first stage (A–B), gravity has no effect and the body moves in the direction of the impetus imparted by the mover. In the second stage (B–C), gravity recovers its power and the impetus declines, so the body starts to tend downward. In the third stage (C–D), gravity takes over and pulls the body downward as the impetus is exhausted.

SCURRILOUS RUMORS

The stories of Buridan's life that have come down to us today may not all be true but they do suggest that he was a lively and colorful character. It is said that he hit the future Pope Clement VI over the head with a shoe during a dispute over a woman, and that he died after the King of France threw him into the Seine River in a sack as punishment for having an affair with the queen.

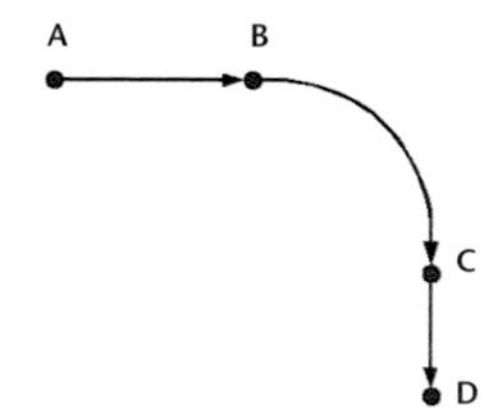

The path of a cannonball shot horizontally from a cannon follows a straight path and then falls to Earth.

The Tunnel Experiment

One of the most important thought experiments in the history of science involves the imagined dropping of a cannonball down a tunnel that goes to the center of the Earth and through to the other side. The experiment was discussed by several medieval thinkers, developing the ideas of Avicenna and Buridan on impetus. The cannonball, it was thought, should rise on the other side of the world to the height from which it was dropped. The explanation was that the cannonball was given impetus by the force of gravity acting on it to pull it into the Earth, and that would be sufficient to counteract gravity on its exit path. When it reached the height from which it was originally dropped, the impetus would be exhausted and the cannonball would fall again, following the same pattern and setting up

"When a mover sets a body in motion he implants into it a certain impetus, that is, a certain force enabling a body to move in the direction in which the mover starts it, be it upward, downward, sideward, or in a circle. The implanted impetus increases in the same ratio as the velocity. It is because of this impetus that a stone moves on after the thrower has ceased moving it. But because of the resistance of the air (and also because of the gravity of the stone) which strives to move it in the opposite direction to the motion caused by the impetus, the latter will weaken all the time. Therefore the motion of the stone will be gradually slower, and finally the impetus is so diminished or destroyed that the gravity of the stone prevails and moves the stone toward its natural place."

Jean Buridan, *Questions on Aristotle's Physics*

THE OXFORD CALCULATORS—CHEATED OF A TRIUMPH

The Oxford Calculators were scientist-mathematicians based at Merton College, Oxford, England, in the 14th century, who included Thomas Bradwardine, William Heytesbury, Richard Swineshead, and John Dumbleton. They investigated instantaneous velocity and came up with the basis of the law of falling bodies long before Galileo, to whom it is usually attributed. They also stated and demonstrated the mean speed theorem: that if a moving object accelerates at a uniform rate for a certain time, it covers the same distance as an object moving at its mean speed for the same period of time. They were among the first to treat properties such as heat and force as theoretically quantifiable, even though they had no way of measuring them, and suggested the use of mathematics in problems of natural philosophy. Unfortunately, medieval Oxford academics were often mocked for the abstruse nature of their studies and the group disappeared into virtual obscurity.

an oscillatory movement. This was the first point at which the concept of oscillatory motion, which was so important in 17th-century physics, was brought into the study of dynamics.

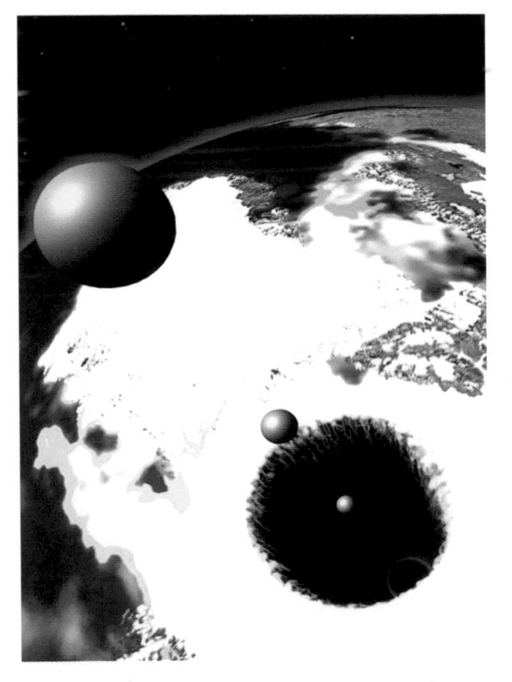

The tunnel experiment was adapted to explain the swing of a pendulum, which was seen as the tunnel experiment in a microcosm. The pendulum is pulled downward to its lowest point (the horizontal mid-point), and the impetus it has gained impels it on its continued lateral (but also upward) path, until that force is exhausted and it is pulled back, renewing the impetus but in the other direction. For Aristotelian dynamics, and in the models of Hipparchus and Philoponus, the pendulum had been an inexplicable anomaly. There was no obvious reason for it to rise again after it had fallen. Here, at last, was a way of explaining it.

The True Birth of Classical Mechanics

The scientists of the 16th and 17th centuries sought explanations for the movement of physical bodies ranging from

A famous thought-experiment involves dropping a cannonball straight through the Earth.

projectiles to stars. The early work on dynamics was examined rigorously and superseded, principally through the efforts of Galileo in Italy and Isaac Newton in England, though with important contributions from astronomers such as Johannes Kepler.

Galileo's Rolling Ball Experiment

Galileo's distrust of Aristotelian physics began at an early age. Even in his teens as a student at Pisa he could refute Aristotle's claim that heavy bodies fall more quickly than lighter ones, citing as evidence that hailstones of different sizes reach the ground at the same time and presumably fall from the same height. (It is a spurious proof, of course, as he had no way of knowing if the hailstones started their fall at the same time.) He showed, too, that a cannonball that hits a target at the same height as it left the cannon does so at the same speed at which it left the cannon.

Galileo had a particular interest in projectiles and falling bodies. It's unlikely that he ever actually carried out the famous experiment attributed to him of dropping cannonballs of different weights from the Leaning Tower of Pisa to show that they fall at the same speed—it's more likely that this was simply a thought experiment. But whether or not he did it, the concept of conducting an experiment to test an idea, and of using the results as evidence

DESCARTES AND THE MECHANISTIC VIEW

René Descartes was, essentially, the first person to propose that there are immutable laws of nature. He developed a mechanistic view, inspired by an amateur scientist and champion of the mechanical philosophy, the Dutchman Isaac Beeckman (1588–1637), whom he met in 1618. Descartes attempted to explain all of the material world, including organic life, in terms of the size, shape, and action of particles of matter moving according to physical laws. He saw even the human body as a type of machine, though the soul was excluded from his mechanistic schema. In his view, God was the prime mover who gave the universe the shove it needed to get it going, but thereafter it ran on its own, following the laws of physics like a piece of clockwork. He believed that if the initial conditions were known, the outcome of any system could be predicted.

Descartes believed that animate beings work like clockwork, following physical laws.

It is unlikely that Galileo ever dropped cannonballs from the Leaning Tower of Pisa, but the idea has had an enduring appeal.

to support a scientific statement, was central to Galileo's practices and would become the basis of the scientific method.

Rather than dropping cannonballs from a dangerous height, Galileo carried out his experiments with forces by rolling balls of different weights down slopes. At a time before watches had a second hand, timing things accurately in experiments was not easy. Galileo used a water clock and his own pulse to measure the time the balls took to reach the end of the slope, showing that the effects of gravity were the same on both light and heavy objects. This was counter to Aristotle's teachings and—apparently—to common sense. But Galileo pointed out that when we see a feather or a sheet of paper fall more slowly than a cannonball, it is because air resistance slows its fall, not because gravity has less influence on the lighter object.

The rolling-ball experiment showed something else, too. As he made the slopes shallower and shallower, Galileo realized that, without a force to stop it, a ball rolled along a horizontal plane would continue rolling forever. Again this went against Aristotle's teachings. It would also appear to be counter-intuitive—a brick pushed along a table stops as soon as we stop pushing it, and even a wheeled trolley will stop after a while. Galileo correctly identified a force that acts to stop the movement—friction. However, he made one error in interpreting his finding that movement will continue unless prevented; he assumed that, since the Earth is turning, inertial motion would always produce a circular path. It was left for Descartes to demonstrate that moving objects continue in a straight line unless some force acts on them to change the direction of their travel.

Stopping and Starting

Inertia is the reluctance of a body to begin moving. It must be overcome if movement is to start. Momentum is the tendency of a moving body to keep moving once it has

GALILEO'S EXPERIMENT ON THE MOON

In 1971, astronauts from Apollo 15 demonstrated that what Galileo had said about falling bodies was correct. When there is no atmosphere (and so no air resistance or lift), falling objects dropped at the same time from the same height hit the ground simultaneously, regardless of their weight or shape. The astronauts used a feather and a geological hammer to demonstrate this.

GALILEO GALILEI (1564–1642)

Galileo was educated at home until the age of 11, when he was sent to a monastery for more formal instruction. To his father's horror, Galileo took to the monastic life and at 15 decided to become a novice monk. Luckily for the history of science, he developed an eye infection and his father brought him home to Florence for treatment. Galileo never returned to the monastery. At his father's prompting, Galileo went to university in Pisa to study medicine, but was quickly distracted by mathematics and did little work on his medical courses. He left without a degree in 1585, but returned four years later as professor of mathematics.

Galileo was poorly paid as a professor and his poverty was exacerbated when his father died having promised (but not delivered) a large dowry for Galileo's sister. He managed to acquire the post of professor of mathematics at Padua in 1592, a more prestigious university and a better-paid job. He still had money worries, though, and turned to invention as a way of alleviating them, developing first a commercially unsuccessful thermometer and then a mechanical calculating machine that did bring in an income for a while. In 1604, Galileo worked with Kepler examining a new star (actually a supernova), and around 1608 he demonstrated that the path of a projectile is parabolic. In 1609 Galileo began making his own telescopes, and over the course of that year had improved the power from three to twenty times magnification compared with the existing design. He sent an instrument to Kepler who used it to confirm Galileo's discoveries in astronomy. These discoveries, such as the moons of Jupiter and the phases of Venus (see page 161), supported the Copernican view that the Earth travels around the sun (heliocentrism), rather than the sun traveling around the Earth (geocentrism). Galileo was for many years guarded in expressing or publishing this view as it ran counter to the doctrine of the Catholic Church and in 1616 he was forbidden to promote or teach the heliocentric model. In 1632 he was given permission to publish a balanced discussion of the subject called *Dialogue Concerning the Two Chief World Systems,* but it was so clearly biased against geocentrism that Galileo was convicted of heresy in 1634 and spent the rest of his life under house arrest. During his seclusion he finished his *Discourses and Mathematical Demonstrations Concerning Two New Sciences,* in which he spelled out the scientific method and stated that the universe could be understood by the human intellect and is governed by laws that can be reduced to mathematics.

spin thread was powered by running water. The first steam-powered devices were pumps, but with James Watt's greatly improved steam engine the power of steam could be used to do many different kinds of work. These inventions were not the work of physicists, but of practical people who needed to achieve a practical task and looked for a practical solution. These solutions came about through observation and inspiration rather than theorizing. Science soon stepped in to help explain and improve the machinery of the Industrial Revolution, and has done so ever since.

Putting Newtonian Mechanics on a New Footing

Newton's laws laid the foundations for classical mechanics, but were extended and developed over the ensuing centuries. The Swiss mathematician and scientist Leonard Euler (1707–83) expanded the scope of Newton's laws from particles to rigid bodies (idealized solid bodies of finite size) and gave two further laws that explain that the internal forces within a body need not be equally distributed. Euler's principle of least action (that nature is lazy) has many applications in physics—notably that light follows the shortest path. The brilliant Italian-French mathematician Joesph-Louis Lagrange (1736–1813) succeeded Euler as director of the Berlin Academy of Sciences. He helped to bring together all developments in Newtonian mechanics in the century after Newton's death and reformulated them into Lagrangian mechanics. In *Méchanique analytique (Analytical mechanics)*, which he began when 19 and finished at the age of 52, Lagrange presented a synthesis of all that had passed in the intervening years based on his own mathematical system, which described the limits of a mechanical system in terms of all the variations that could happen over the course of its history expressed using calculus. The Lagrangian equations relate the kinetic energy of a system to its generalized coordinates, generalized forces, and time. His book contains no diagrams—a notable achievement for a book on mechanics; his methods use calculus and exclude geometry. His work simplified many calculations in dynamics by dealing with scalar functions of kinetic and potential energy rather than an accumulation of forces, accelerations, and other vector quantities.

Both Euler and Lagrange tackled fluid dynamics, too, but took different approaches. Euler described the movement of particular points in a fluid, while Lagrange divided the fluid into regions and analyzed their trajectories.

Another mathematician who made significant contributions to modern practical mechanics was the Irish nobleman Sir William Rowan Hamilton (1805–65). In his treatise *On a General Method in Dynamics* (1835) he expressed the energy of a system in terms of momentum and position, reducing dynamics to a problem in the calculus of variations. His reformulation of classical mechanics in the Hamiltonian equations is sometimes called Hamiltonian mechanics. In the process he discovered that there is a close link between Newtonian mechanics and geometric optics. The full significance of his work was not apparent until the rise of quantum mechanics nearly a hundred years later.

Inertia and Gravity Come Together

Between Newton's statement of the laws of inertia and gravity and Einstein's relativity theories came the Austrian physicist Ernst Mach (1838–1916). Newton believed space to be an absolute backdrop against which motion can be plotted. Mach disagreed, saying that movement is always relative to another object or point. Like Einstein, he believed only relative motion makes sense. As a consequence, inertia can only be understood if there are other objects against which we can compare the movement or stillness of a body. If there were no stars or planets, for example, we would not be able to tell that Earth was spinning. Mach's principle—which he did not present as a principle himself, it was Einstein who coined the term—has been stated in rather general terms as something like "mass there influences inertia here." With no mass "there," there can be no inertia "here."

SIR WILLIAM ROWAN HAMILTON (1805–65)

Brilliant from early childhood, Hamilton learned to read at the age of three. He could translate Latin, Greek, and Hebrew by the age of five, compiled a grammar of Syriac at 11, and at 14 composed a welcome in Persian to the Persian ambassador who was visiting Dublin. Hamilton's gift for mathematics and astronomy was so considerable that he was elected professor of astronomy and Astronomer Royal in Ireland while still an undergraduate. He relied rather heavily on alcohol as a source of sustenance, and although he carried out most of his work in the dining room of his home, he ate little other than mutton chops. Dozens of plates with mutton bones still in place were found among many of his papers after his death. His achievements spanned mathematics, astronomy, classics, dynamics, optics, and mechanics.

Big and Small

While Newtonian mechanics seemed to work well for the larger objects in the universe, it began to fall down when applied to the very tiniest. As physicists became aware of atomic and subatomic particles, they discovered that the laws of physics, which they considered to be fixed and immutable for all things, did not seem to apply any more. The smallest particles could do odd things. The hard-won confidence in the laws of physics was foundering, and in the 20th century these laws were to come under close scrutiny.

Instead of work on the atom demonstrating that Newton's ideas really did explain the entire universe, it showed that at very small scales, matter behaves in very surprising ways. Classical mechanics reaches its limits at the atomic scale, at velocities near the speed of light, and in intense gravitational fields. Before looking into the atom and how it seems to defy the laws of nature, we need to backtrack a little and examine energy—the other half of the mass-in-motion equation.

CHAPTER 4

ENERGY
Fields and Forces

When a force acts to move a mass it seems obvious to us that energy is involved. So it may seem surprising that for all the consideration of forces since antiquity, energy was largely neglected by the early natural philosophers. The concept of energy is relatively new, emerging only in the 17th century. Indeed, the term "energy" (from the Greek *energia*, coined by Aristotle) was only introduced with its modern meaning in 1807 by the genius and polymath Thomas Young (of the double-slit experiment). The most obvious forms of energy are light and heat, both of which come for free from the sun. Humankind has also harnessed chemical energy (released by burning fuels), the gravitational energy of a falling body, the kinetic energy of wind and moving water, and, latterly, electric and nuclear energy.

Lightning and wind represent massive bursts of energy in nature, feared for their destructive power.

The Conservation of Energy

Just as matter is conserved, being neither created nor destroyed, so energy too is conserved. It may be converted from one form to another—and this is how we harness energy to do useful work—but that energy is never actually spent. Galileo noticed that a pendulum converts gravitational potential energy into kinetic energy or the energy of movement. When the pendulum bob is at the high point of its swing it is momentarily still, and has maximum potential energy. This is converted into kinetic energy as the bob moves, and the bob regains potential energy as it climbs up on the other side of its swing.

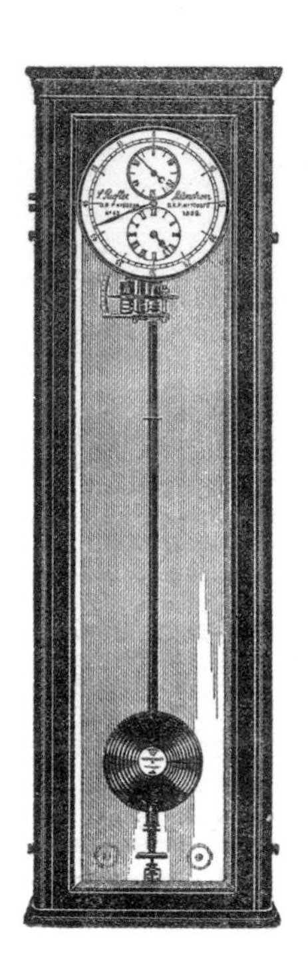

Pendulum clocks were first developed in 1656 by Christiaan Huygens: the pendulum always takes the same time to swing.

INVENTING "ENERGY"

That different types of energy are equivalent was not immediately obvious. Even now there is no fundamental understanding of exactly what energy is and how it works. The German mathematician

> *"There is a fact, or if you wish, a* law, *governing all natural phenomena that are known to date. There is no known exception to this law—it is exact so far as we know. The law is called the* conservation of energy. *It states that there is a certain quantity, which we call energy, that does not change in manifold changes which nature undergoes. That is a most abstract idea, because it is a mathematical principle; it says that there is a numerical quantity which does not change when something happens. It is not a description of a mechanism, or anything concrete; it is just a strange fact that we can calculate some number and when we finish watching nature go through her tricks and calculate the number again, it is the same."*
>
> US physicist Richard Feynmann, 1961

An ice skater can speed up her turns by holding her arms close to her body, or slow them down by spreading her limbs.

Gottfried Leibnitz (1646–1716) explained mathematically the conversion between different types of energy, which he called *vis viva*. His work, along with observations by the Dutch mathematician and philosopher Willem Gravesande (1688–1742), was refined by the French physicist Marquise Émilie du Châtelet (1706–49), who defined the energy of a moving body as proportional to its mass multiplied by its velocity squared. The current definition of kinetic energy is very close to this:

$$E_k = \frac{1}{2} mv^2$$

Humans used fire for thousands of years with no understanding of how it works.

Struggling with Fire

Early theories about how and why things burn centered around a supposed component of flammable matter called phlogiston. When the material was burned, the phlogiston escaped. This was not really a theory of energy but of the physical and chemical changes brought about by fire. The theory originated in 1667 with the work of the alchemist Johann Becher (1635–82). He revised the ancient model of matter comprising four elements—earth, air, water, and fire—dating from the days of Empedocles (see page 20) and replaced it with three forms of earth: *terra lapidea*, *terra fluida*, and *terra pinguis*. In 1703, Georg

PERPETUAL MOTION MACHINES

The principle of the conservation of energy might suggest that it's possible to make a perpetual motion machine: one that uses the energy it produces to keep itself going, so constantly recycling its energy between different forms. The idea was first suggested around 1150 by the Indian mathematician Bhaskara (1114–85) who described a wheel that dropped weights along the length of its spokes as it rolled, so propelling itself along. Even Robert Boyle, who might have been expected to know better, proposed a system that continually filled a cup with water, emptied it, and refilled it. All ideas for perpetual motion machines must fail, however, as energy is lost to friction and through inefficiency. In the 18th century, both the French Royal Academy of Sciences and the American Patent Office were so overwhelmed with applications and proposals for perpetual motion machines that they banned them.

Lavoisier's laboratory in Paris

Ernst Stahl (1660–1734), the professor of medicine and chemistry at the university of Halle in Germany, changed the model slightly and renamed *terra pinguis* "phlogiston."

Phlogiston was held to be a substance without odor, color, or taste that is freed by matter when it is burned. When all the phlogiston has been released, the nature of the burnt matter is usually different, as when wood turns to ash. However, if matter is burned in an enclosed space, it may not all burn away as the air becomes saturated with phlogiston. There were difficulties in explaining how sometimes metals increase in mass when burned or heated (now known to be because they form oxides), but phlogiston theorists had a cunning solution to this. They claimed that sometimes phlogiston has no weight, sometimes it has positive weight, and sometimes it even has negative weight, so the loss of phlogiston can actually increase the mass of the burned matter. Phlogiston was also implicated in rusting and in living systems—a creature cannot live in "phlogisticated" air in which something has been burned, nor will iron rust in it.

The theory was not overturned in favor of a chemical explanation until Antoine-Laurent Lavoisier (see page 30) demonstrated that when material burns or rusts it combines with oxygen. The realization that this had a link with living processes—that respiration also requires oxygen—was the first clue that chemical processes are at the heart of life.

While phlogiston and then oxygen explained the chemical process of burning, heat itself remained a mystery until 1737 when du Châtelet proposed what was later recognized as infrared radiation.

Georg Ernst Stahl

GABRIELLE ÉMILIE LE TONNELIER DE BRETEUIL, MARQUISE DU CHÂTELET (1706–49)

The daughter of a French aristocrat, Émilie du Châtelet was considered too tall for a woman and so her father thought she was unlikely to marry. As a consequence, he employed the best tutors for her (she could speak six languages by the age of 12), and allowed her to indulge her interest in physics and maths. Her mother disapproved and wanted to send her to a convent, but fortunately her father's view prevailed. Émilie developed an interest in gambling, using mathematics to improve her chances of winning, then used her winnings to buy books and laboratory equipment.

Émilie did marry, and had three children. As her husband was often away on military campaigns or visiting his numerous estates, she was free to carry on with her scientific pursuits and take lovers—probably including writer and philosopher Voltaire, whose real name was François-Marie Arouet. He was certainly her close intellectual companion and spent a lot of time in the du Châtelet estate at Cirey-sur-Blaise where the pair shared a laboratory. Émilie translated Newton's *Principia*, and wrote *Institutions de physiques* (1740), which tried to reconcile Newton's and Leibnitz's views. In 1737 she entered a competition run by the *Académie de science* with a paper she researched in secret on the properties of fire. In this she suggested that different colors of light had different power to heat, foreshadowing the identification of infrared radiation. She did not win the competition, but her paper was published.

One of her experiments involved dropping cannonballs into a bed of wet clay. She found that doubling the velocity of the cannonball caused it to go four times as deeply into the clay, showing that force is proportional to mass times velocity squared ($m \times v^2$) and not, as Newton said, mass times velocity.

Émilie du Châtelet was a notable female physicist at a time when science was a male preserve.

One of Rumford's experiments with cannon barrels. He proposed that heat is moving particles, and can be caused by friction.

Thermodynamics

The development of the steam engine and many other powered machines of the industrial revolution meant there was an increasingly urgent need to understand thermodynamics—how heat is produced, transferred, and can be harnessed to do physical work. Two theories about the nature of heat, not quite mutually exclusive but strange bedfellows, were current in the 18th century: the caloric model and the mechanical model of heat.

The mechanical model is based on the movement of tiny particles. The kinetic theory of gases has its origins in Daniel Bernoulli's book *Hydrodynamica*, published in 1738. He suggested that gases are made up of moving molecules. When they bombard a surface, the effect is pressure; their kinetic energy is felt as heat. This is the model that is still accepted today.

The caloric model suggested that heat is a form of matter, a sort of gas with indestructible particles. Atoms of heat—or caloric—could combine with the atoms of other substances or could be free-ranging and sneak between the atoms in other matter. Lavoisier proposed the existence of

"I am now as much convinced of the non-existence of caloric as I am of the existence of light."

Humphry Davy, 1799

FROZEN OUT

Just as heat was supposed to be the result of caloric, some scientists of the 1780s believed cold to be a property produced by the presence of a substance called "frigoric." This was discredited by the Swiss philosopher and physicist Pierre Prévost (1751–1839), who said that cold is simply an absence of heat, and showed in 1791 that all bodies, no matter how cold they appear to be, radiate some heat.

caloric while debunking phlogiston. He believed that caloric atoms were a constituent of oxygen and their release produced the heat of combustion. When heat was produced by friction, this occurred because atoms of caloric were rubbed off the moving body.

The American-born physicist Benjamin Thompson Count Rumford (1753–1814) conducted an experiment in which he weighed ice, melted it, and then weighed it again. He found that there was no discernible difference in weight, suggesting there was no caloric gained in melting the ice. But supporters of the caloric model countered by suggesting instead that caloric had negligible mass. Count Rumford's additional observation that the act of boring holes in metal, such as cannon barrels, produced an enormous amount of heat, together with experiments carried out by the English chemist Humphry Davy (1778–1829), should have demonstrated to everyone that the caloric theory was wrong, as they showed that heat could be produced by physical work alone. Although some people doubted the caloric theory, the conclusions of Count Rumford and Davy were not accepted until the English physicist James Prescott Joule (1818–89) repeated some of their experiments 50 years later.

Joule carried out experiments to demonstrate that work could be converted into heat. For example, forcing water at pressure through a perforated cylinder raises the temperature of the water. This laid the foundations for the theory of the conservation of energy through its transfer into different forms, and showed that the caloric model of heat was not correct. (Strangely, the conservation of heat energy was integral to the caloric model as it made heat into matter, which was already known to be conserved.)

Joule calculated that the amount of work needed to raise the temperature of one pound of water by one degree Fahrenheit was 838 foot-pounds force. (One foot-pound is the torque—or twisting force—created by a one-pound force acting at a

Joule's equipment measured the mechanical equivalent of heat.

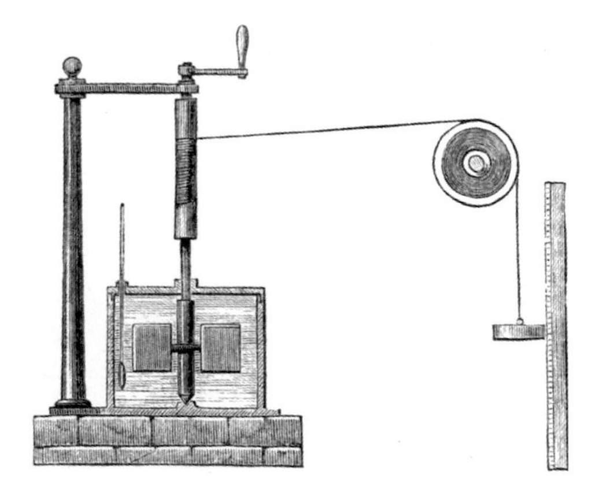

PHYSICS GETS STATISTICAL

James Clerk Maxwell's formulation of molecular velocities, known as the Maxwell distribution, provided a way of calculating the proportion of molecules with a specific velocity (or the probability of a particle having a specific velocity) in a gas where molecules have free movement. It was the first statistical law in physics. It has been replaced by the Maxwell-Boltzmann distribution, which refines Maxwell's technique and assumptions.

perpendicular distance of one foot from a pivot point.) He tried different methods and got similar results, leading him to accept that his theory and his figures were approximately correct.

Joule's work received an unenthusiastic reception at first, partly because it relied on very precise measurements—differences in temperature of $\frac{1}{200}$th of a degree.

When Michael Faraday and William Thomson (later Lord Kelvin) heard Joule's presentation of his work in 1847 they were both interested, but it took a long time for them to share his view.

The first collaboration with Thomson took place when the two met while Joule was on honeymoon. They planned to measure the difference in temperature of the water at the top and bottom of a waterfall in France, but in the end it proved impractical. Thomson and Joule corresponded from 1852 to 1856, with Joule carrying out experiments and Thomson commenting on the results. Joule concluded that heat is a form of movement of atoms. Although the atomic model of matter was not universally accepted at this point, Joule had learned all about it from the English chemist John Dalton (see page 31) and accepted the atomic model wholeheartedly.

The Laws of Thermodynamics

Three laws of thermodynamics set the limits on what can and cannot be done in any system involving heat and energy. The laws emerged during the 19th century, once heat became generally accepted as the movement of particles.

The first law of thermodynamics, formulated by Rudolf Clausius (1822–88) in 1850, is essentially a statement of the conservation of energy: the change in the internal energy of a system is equal to the amount of heat supplied to it, less the amount of work performed by the system. In other words, energy is never created or destroyed. The law, as stated by Clausius, was based on Joule's demonstration that work (or energy) is equivalent to heat.

The second law of thermodynamics was actually discovered before the first. The

A steam engine converts heat energy into kinetic energy to drive a vehicle or machinery.

NICOLAS LÉONARD SADI CARNOT (1796–1832)

Born in Paris, France, Nicolas Carnot was the son of a military leader and cousin of Marie François Sadi Carnot (president of the French Republic 1887–94). From 1812 the young Carnot attended the École polytechnique in Paris where he was probably taught by the noted physicists Siméon-Denis Poisson (1781–1840), Joseph Louis Gay-Lussac (1778–1850), and André-Marie Ampère (1775–1836). The steam engine, in use since 1712, had been greatly improved by James Watt more than fifty years later. Yet its development had been largely a matter of trial and error and inspired guesswork but little scientific study. At the time that Carnot began investigating the steam engine, it had an average efficiency of only 3 percent. He set out to answer two questions: "Is the work available from a heat source potentially unbounded?" and "Can heat engines be improved by replacing the steam with some other working fluid or gas?" In tackling these questions, he came up with a mathematical model of the steam engine that helped scientists to understand how it worked.

Although Carnot stated his findings in terms of caloric, his work laid the foundations for the second law of thermodynamics. He found that a steam engine produces power not because of the "consumption of caloric, but [because of] its transportation from a warm body to a cold body" and that the power produced increases with the difference in temperature "between the warm and cold bodies." He published his conclusions in 1824, but his work earned little recognition until resurrected by Rudolf Clausius in 1850.

Carnot died of cholera at the age of only 36. Because of concerns about infection, most of his papers and other belongings were buried with him, leaving only his book as testament to his work.

Nicolas Sadi Carnot

French military engineer Nicholas Sadi Carnot (see box, above) described a theoretical ideal heat machine in which no energy is lost to friction or wastage and demonstrated that the efficiency of the machine depends on the difference in temperature between the two bodies. So a steam engine using super-heated steam will produce more work than one using cooler steam and, eventually, an engine (such as diesel) that uses fuel at a higher temperature will be more efficient still. Like much of the

work on thermodynamics in the 19th century, Carnot took the design of existing machinery as his starting point to explore and explain the physics that made it work. Practical science was driving theoretical science.

Carnot stated his findings in terms of caloric, and it was Clausius who restated the law in terms of entropy, saying that a system always tends toward a greater state of entropy. Entropy is commonly considered to be "disorder." More precisely, it is a measure of the unavailability of energy in a system to do work; in any real system some energy is always lost in dissipated heat. When fuel is burned, the energy is converted from an organized (low-entropy) state to a disorganized (high-entropy) state. The total entropy of the universe increases every time fuel is burned. Clausius summarized the first and second laws by saying that the amount of energy in the universe remains constant but its entropy tends toward a maximum. The end of the universe, if this is taken to its extreme, will be a vast soup of dissociated atoms. This situation, called heat-death of the universe, was first proposed by Clausius.

The third law of thermodynamics followed much later, in 1912. Developed by the German physicist and chemist Walther Nernst (1864–1941), it states that no system can reach absolute zero, the temperature at which atomic movement almost ceases and entropy tends to a minimum or zero.

MAXWELL'S DEMON

In 1871, James Clerk Maxwell proposed a thought experiment to try to cheat the second law of thermodynamics. He described two adjacent boxes, one containing hot gas and one containing cold gas, with a small hole connecting them. Normally, heat moves from a hot area to a cold area, fast particles bash into slow particles and speed them up and vice-versa. Eventually the gas in both boxes will contain particles with a similar distribution of speeds and will be the same temperature. In the experiment, though, a demon sits at the hole regulating the particles that can pass through. The demon opens the hole to allow fast-moving particles from the box of cold gas into the box of hot gas and slow-moving particles from the box of cold gas into the box of hot gas. In this way the demon can raise the temperature of the hot gas at the expense of the cold gas and decrease the entropy of the system. The system still cannot cheat the law, as anything performing the function of the demon must itself use energy to work. In 2007, the Scottish physicist David Leigh made a nanoscale attempt at a demon machine. It can separate slow- and fast-moving particles, but needs a power supply of its own.

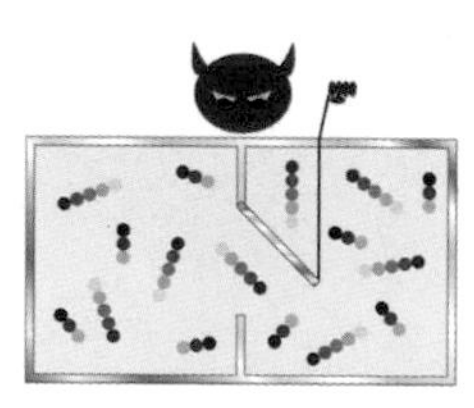

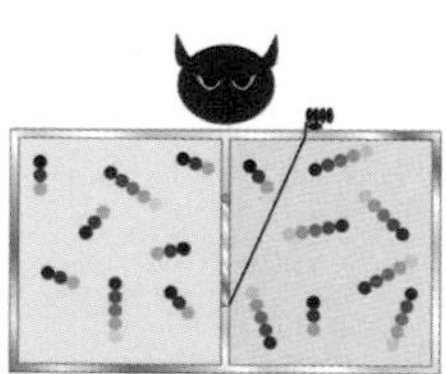

ABSOLUTE ZERO

The third law of thermodynamics requires the concept of a minimum temperature below which no temperature can ever fall—known as absolute zero. Robert Boyle first discussed the concept of a minimum possible temperature in 1665 in *New Experiments and Observations Touching Cold*, in which he referred to the idea as *primum frigidum*. Many scientists of the time believed there was "some body or other that is of its own nature supremely cold and by participation of which all other bodies obtain that quality."

French physicist Guillaume Amontons (1663–1705) was the first to tackle the problem practically. In 1702 he constructed an air thermometer, and declared that the temperature at which air had no "spring" to affect the measurement was "absolute zero." Zero on his scale was around –400° Fahrenheit (–240° Celsius). The Swiss mathematician and physicist Johann Heinrich Lambert (1728–77), who proposed an absolute temperature scale in 1777, refined the figure to –454°F (–270°C) —close to the currently accepted figure.

This nearly correct figure wasn't universally accepted, though. Pierre-Simon Laplace and Antoine Lavoisier suggested in 1780 that absolute zero may be 1,500 to 3,000 degrees below the freezing-point of water, and that at the very least it must be 600 degrees below freezing. John Dalton gave the figure of –5432°F (–3000°C). Joseph Gay-Lussac came closer after his investigations of how the volume and temperature of a gas are related. He found that if the pressure is kept constant, the volume of a gas increases by 1/273 for each rise of 33.8°F (1°C) above zero.

A Galileo thermometer depends on the variation in pressure with temperature; at absolute zero, no pressure is exerted as atoms do not move.

From this he could extrapolate backward to a figure for absolute zero of –459.4°F (–273°C)—even closer to the correct figure.

The problem took a different turn after Joule had shown that heat is mechanical. In 1848, William Thomson (later Lord Kelvin) devised a temperature scale based on the laws of thermodynamics alone, not on the properties of any particular substance (unlike Fahrenheit and Celsius). Kelvin found a value for absolute zero that is still accepted, –459.67°F (–273.15°C)—very close to the value derived from the air thermometer and Gay-Lussac's theory. The Kelvin scale is based on the Celsius scale but starts at –459.67°F, rather than 32°F (0°C). Although hugely influential, knighted, and appointed President of the Royal Society, Kelvin was not an entirely discerning scientist, and rejected both Darwin's theory of evolution and the existence of atoms.

Heat and Light

It has been clear to humankind for millennia that sunlight provides both light and heat, but the link between them has been explained only relatively recently. The first person known to note the connection

HOW COLD?

Even outer space is not at absolute zero. The ambient temperature in outer space is 2.7 Kelvin as cosmic microwave background radiation—heat left over from the Big Bang—is present throughout space. The coldest known area is in the Boomerang Nebula, a dark gas cloud at just over 1 Kelvin. The lowest temperature ever achieved artificially is 0.5 billionths of a Kelvin attained briefly in a laboratory at Massachusetts Institute of Technology (MIT) in 2003.

was an Italian scholar, Giambattista della Porta (*ca.*1535–1615) who, in 1606, noted the heating effect of light. Something of a polymath, della Porta was a playwright as well as a scientist, who published on agriculture, chemistry, physics, and mathematics. His *Magiae naturalis* (1558) inspired the foundation of the Italian scientific academy, the Lyncean Academy (Accademia dei Lincei or "Academy of the Lynxes"), in 1603. (The title page of the book was illustrated with a lynx design, and the prologue contained a description of the scientist as one who "with lynx-like eyes, examin[es] those things which manifest themselves, so that having observed them, he may zealously use them.")

Émilie du Châtelet drew a link between heat and light when she noticed that the heating power of light varied with its color. Although this looked ahead to the electromagnetic spectrum and the discovery of infrared radiation, it was not developed further at the time. In 1901 Max Planck (see panel, below) made an important discovery linking light and heat while researching black-body radiation, but his was an accidental breakthrough, the result of a fudge. That fudge, though, was to form the basis of quantum mechanics.

"[The quantum solution to the black-body problem] was an act of despair because a theoretical interpretation had to be found at any price, no matter how high that [price] might be"

Max Planck, 1901

Black-Body Radiation and Energy Quanta

Many types of material glow as they are heated up, emitting light that moves from red through yellow to white. The wavelength of the light

The Accademia dei Lincei has occupied the Palazzo Corsini in Rome since 1883.

emitted at higher temperatures becomes progressively shorter, as it moves toward the blue end of the spectrum. As this is added to the yellow and red light, the glow of the hot body becomes whiter and then bluer. The graph that shows this distribution of heat and color is called the black-body curve. The perfect "black body" is something that absorbs all the radiation that falls on to it. A box made of graphite with a tiny hole in it is a good approximation of a perfect black body (the hole acts as the black body). When the black body is heated, it glows, radiating light at different wavelengths for different temperatures. The color of the radiated light depends entirely on the temperature, not the material, of the body.

Planck tried to calculate the exact amount of light emitted at different wavelengths by a black body that consisted of a black box with a tiny hole in it. Although he could very nearly get his equation to give a correct result, he had to make an odd assumption to make it perfect. That assumption was that instead of light coming from the box in a constant stream, as he would expect the wave to be, it had to be cut into discontinuous little chunks or packets of wave—or quanta. Planck had not actually intended that energy quanta would become part of the landscape of physics. He saw them instead as a neat mathematical trick that would be replaced at some point in the future by a new discovery or calculation. How wrong he was.

Other Forms of Energy

While light and heat were coming under scrutiny, some new forms of energy were also coming to the attention of the scientific community. Many types of energy that had been exploited for years were only named in the 19th century. The French scientist Gustave-Gaspard de Coriolis (1792–1843)

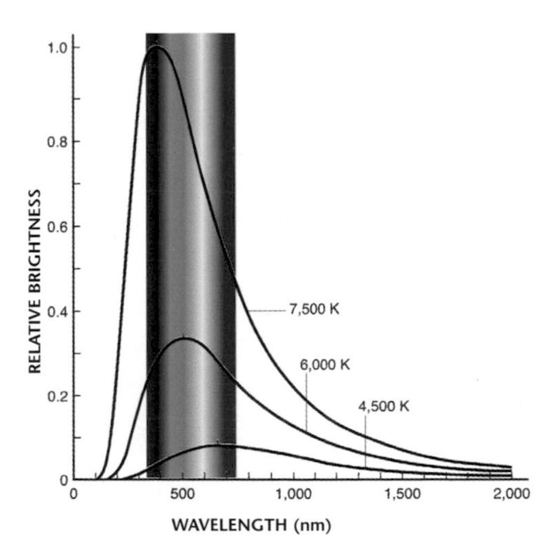

At 7500 K the black body radiates light at the violet end of the spectrum; at 4500 K it has shifted toward the red.

Spectroscopy, working from the light emitted by glowing lava, can be used to calculate the temperature of the lava flow from a volcanic eruption.

described kinetic energy in 1829, and the term "potential energy" was coined by the Scottish physicist William Rankine (1820–72) in 1853. First among the newly recognized sources of energy was electricity. Although lightning was familiar to everyone, no-one had realized that it involved electricity.

Discovering Electricity

The first type to be discovered was static electricity. Even in ancient times, people were aware that rubbing amber or jet created a kind of force that made the material attract fluff and bits of material, but the nature of the attraction was not understood. The

MAX PLANCK (1858–1947)

Max Planck had a long but tragic life. Born in Kiel, in the Duchy of Holstein (now Germany), he first wanted to be a musician. When he asked another musician what he should study, the man told him that if he needed to ask he wouldn't make a musician. He then turned his attention to physics, only for his physics professor to tell him that there was nothing left to discover. Fortunately, Planck stuck with it and his formulation of quanta laid the groundwork for much of 20th-century physics.

Planck's first wife died in 1909, possibly of tuberculosis. During the First World War, one of his sons was killed on the Western Front, and another, Erwin, was taken prisoner by the French. Planck's daughter Grete died in childbirth in 1917, and her twin sister Emma died the same way in 1919 (after marrying Grete's widower). In 1944, Planck's house in Berlin was entirely destroyed during an Allied bombing raid, and his scientific papers and correspondence were all lost. The final straw came in 1945 when Erwin was executed by the Nazis for colluding in a plot to kill Hitler. Planck lost the will to live after Erwin's execution, and died in 1947.

Otto von Guericke's electricity generator worked with static electricity.

English natural philosopher Sir Thomas Browne (1605–82) defined "electric" as "a power to attract strawes and light bodies, and convert the needle freely placed." In 1663, the German scientist Otto von Guericke built the first electrostatic generator. Guericke had already carried out experiments with air pressure that showed the possibility of a vacuum (see page 27). His electrostatic generator—or "friction machine"—used a globe of sulfur that could be rotated and rubbed by hand to generate a charge. Isaac Newton suggested using a glass globe instead of a sulfur one, and later designs used other materials. In 1746, a friction machine with a large wheel turning several glass globes used a sword and a gun barrel suspended from silk cords as conductors; another used a leather cushion in place of a hand, and one made in 1785 involved two cylinders covered in hare fur that were rubbed together.

Experiments with electricity became more common in the 18th century, and generators of static electricity were popular attractions at public science lectures. Two people, a Dutch mathematics teacher named Pieter van Musschenbroek (1692–1761) and the German cleric Ewald Georg von Kleist (1700–48), independently invented the Leyden jar around 1744. Comprising a jar part-filled with water, with a metal rod or wire through the cork, this was a simple device for storing electricity. A more efficient design had metal foil on the outside of the jar.

When von Kleist first touched his jar, a powerful electric shock knocked him to the floor. The Leyden jar became a valuable tool in experiments with electricity, and is the origin of the modern capacitor. Benjamin Franklin, investigating the jar, discovered that the charge is held in the glass and not, as previously supposed, in the water.

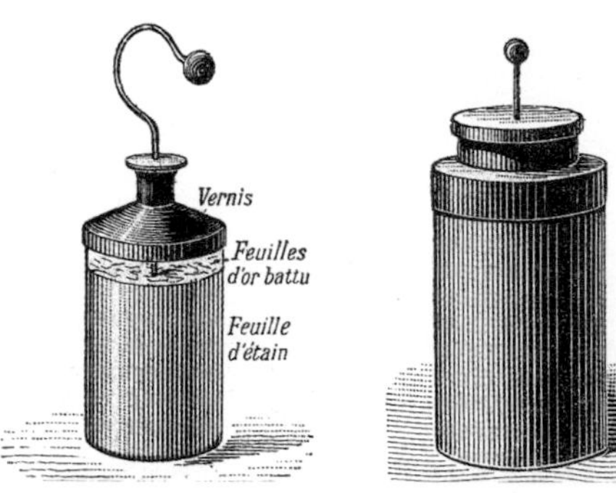

Leyden jars

Kites and Thunderstorms

The American scientist Benjamin Franklin (1706–90), who went on to help draft the American Declaration of Independence, first demonstrated the electrical nature of lightning in 1752. In a famous experiment, he tested his theory by attaching a metal rod

THE FIRST TENS MACHINE?

Ancient Egyptians may have used electric catfish for medical purposes and the Romans certainly found the black torpedo fish useful as a source of pain relief. As the black torpedo fish (*Torpedo torpedo*) produces an electric charge it can be used like a TENS (transcutaneous electrical nerve stimulation) painkilling machine. The Romans used the fish to relieve the pain of gout, headache, surgical operations, and childbirth. The fish did not survive the procedure (presumably because it was used out of water). Attempts to mimic the effect of the electric fish had their highpoint in the leather torpedo fish made by Henry Cavendish in 1776. After studying the fish, he first made a wooden copy but found it did not conduct electricity well. His second dummy fish was made of thick pieces of sheep leather with thin pewter plates on either side to simulate the electrical organs. He connected the plates to Leyden jars, and soaked the leather fish in salt water. Putting his hand in the water near the fish, he felt a shock similar to those described by people who had felt the effects of a real torpedo fish.

to a kite and tying a key to the other end of the string. He flew the kite during a thunderstorm, with the key dangling near a Leyden jar. Even without lightning, there was enough electrical charge in storm clouds for the wet string to conduct electricity to the key and cause sparks to leap to the Leyden jar. Franklin suggested electricity may have a positive or negative charge. He invented the lightning rod, which carries the electrical charge of a lightning strike safely to earth through a metal conduit, and the lightning bell (see panel, opposite).

Trendy Electricity

Experiments with

Benjamin Franklin carried out experiments with lightning to investigate electricity.

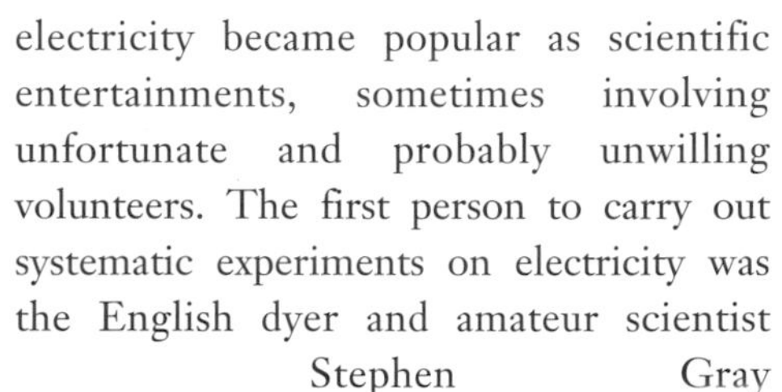

electricity became popular as scientific entertainments, sometimes involving unfortunate and probably unwilling volunteers. The first person to carry out systematic experiments on electricity was the English dyer and amateur scientist Stephen Gray (1666–1736). His "charity boy" was a poor urchin suspended by an insulating cord while holding a charged glass rod, sparks coming from his nose as he attracted tiny particles of metal leaf. Besides being entertaining (at least for the audience), Gray's experiments in 1729 demonstrated conductivity—that electricity could be passed from one material to another, including through water. In a similar experiment, an

"In September 1752, I erected an Iron Rod to draw the Lightning down into my House, in order to make some Experiments on it, with two Bells to give Notice when the Rod should be electrified. A contrivance obvious to every Electrician.

I found the Bells rang sometimes when there was no Lightning or Thunder, but only a dark Cloud over the Rod; that sometimes after a Flash of Lightning they would suddenly stop; and at other times, when they had not rang before, they would, after a Flash, suddenly begin to ring; that the Electricity was sometimes very faint, so that when a small Spark was obtained, another could not be got for sometime after; at other times the Sparks would follow extremely quick, and once I had a continual Stream from Bell to Bell, the size of a Crow-Quill. Even during the same Gust there were considerable variations."

Benjamin Franklin, 1753

electrical charge was passed along a line of old men holding hands. The chemist Charles du Fay (1698–1739), working in Paris, developed Gray's work further, and in 1733 concluded that every object and every living creature contains some electricity. He demonstrated that electricity comes in two forms—negative, which he called "resinous," and positive, or "vitreous." In 1786, the Italian physician Luigi Galvani (1737–98) experimented by passing an electric current through dead frogs, causing their legs to twitch spasmodically. This led him to conclude that the frogs' nerves carry an electrical impulse that causes their leg muscles to work.

Georg Ohm's name is now used for the units of electrical resistance.

Putting Electricity to Work

Before electricity could be put to good use it was necessary to find a way of releasing or generating it as and when it was needed. The first electric cell, the precursor of the battery, was developed by the Italian physicist Alessandro Volta (1745–1827), who gave his name to the volt, the unit of measure for electrical potential. His electric "pile," made in 1800, consisted of a stack of disks of zinc, copper, and paper soaked in salt solution. He had no idea why this produced an electric current but that didn't really matter as it clearly worked. The operation of ions in carrying an electric charge was eventually described in 1884 by the Swedish scientist Svante August Arrhenius (1859–1927). The German physicist Georg Ohm (1789–1854) used a version of Volta's cell for his own investigations of electricity, which led to his formulation of the law that bears his name, published in 1827. Ohm's law states that when electricity is carried through a conductor:

$$I = V/R$$

where I is the current in amps, V is the potential difference in volts, and R is the

resistance in ohms. The resistance of the material remains constant irrespective of the voltage, so changing the voltage directly affects the current.

Waiting in the Wings: Magnetism

We can't get much further with electricity without bringing magnets into the picture. The power of some materials to attract iron, or to line up north–south, was noticed by the ancients but was inexplicable and must have seemed akin to magic.

According to Aristotle, Thales (*ca.*625–545 BCE) gave a description of magnetism in the sixth century BCE. Around 800 BCE the Indian surgeon and writer Sushruta described the use of magnets for removing metal splinters from the body. Another early reference to magnetism is found in a Chinese work written in the fourth century BCE called *Book of the Devil Valley Master*, which says, "The lodestone makes iron come or it attracts it." A lodestone is a naturally magnetized lump of the metal magnetite. Lumps of magnetite with the right crystalline structure might be magnetized by lightning bolts. Chinese fortune-tellers began using lodestones with divining boards during the first century BCE.

A compass uses the Earth's magnetic field to aid navigation.

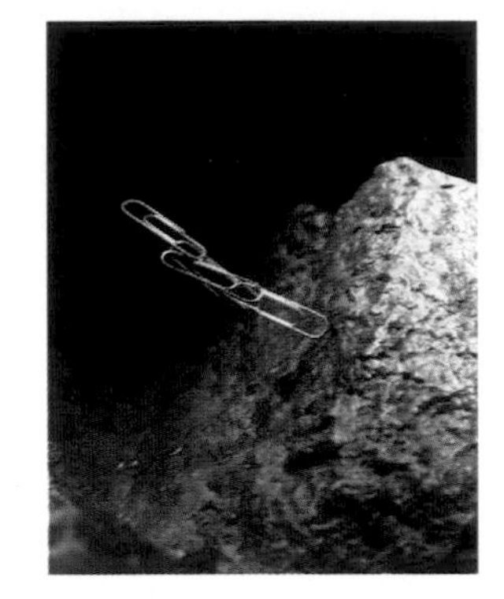

A lodestone is naturally magnetic and will attract magnetic metals such as iron and steel.

Lodestones may possibly have been used in compasses as early as 270, but the first confirmed use of a compass for navigation appeared in Zhu Yu's book *Pingzhou Table Talks* in 1117, which states: "The navigator knows the geography, he watches the stars at night, watches the sun at day; when it is dark and cloudy, he watches the compass." The navigational compass was probably developed independently in Europe. Although the Chinese compass had 24 basic divisions, European types have always had 16. In addition, the compass did not appear in the Middle East until after its first recorded use in Europe, suggesting that it did not pass through the Middle East from China to Europe. Finally, while Chinese compasses were often designed to indicate south, European compasses have always indicated north.

The first scientific investigations of magnetism were carried out by the Englishman William Gilbert (1544–1603), a scientist at the court of Elizabeth I. Gilbert coined the Latin word *electricus*, meaning "of amber." He published his book *De magnete* in 1600, describing in it many experiments he had undertaken to try to discover the nature of magnetism and electricity. It gave the first rational explanation of the mysterious ability of the compass needle to

point north–south, revealing the astounding truth that the Earth is itself magnetic. Gilbert managed to refute the belief popular among sailors that garlic disabled a compass (helmsmen were not allowed to eat garlic near the ship's compass), and the idea that a vast magnetic mountain near the North Pole would draw all the iron nails out of a ship that approached too close to it.

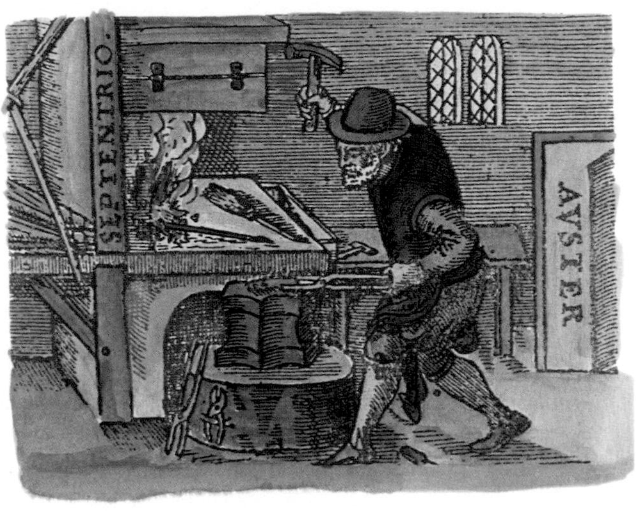

A blacksmith making a magnet, depicted in William Gilbert's De Magnete

The potential power of magnetism was recognized in stories of the iron coffin of Mohammed, which was supposedly kept floating in mid-air by being positioned between two magnets. (Of course, if this spectacle were real, only one magnet above the tomb would have been needed, as gravity would have provided the downward impulse.)

Electromagnetism—The Marriage of Electricity and Magnetism

Practical applications for electricity began to appear in the early 19th century. In 1820, the Danish physicist and chemist Hans Christian Ørsted (1777–1851) noticed that an electric current could deflect the needle of a compass. This was the first hint of a connection between electricity and magnetism. Just a week later, André-Marie Ampère gave a much more detailed account. He demonstrated to the Académie de Science that when parallel wires carry an electric current they either attract or repel each other, depending on whether their currents run in the same or in opposite directions, so laying the foundation for electrodynamics. The following year, Michael Faraday held an experiment in which he placed a magnet in a dish of mercury and suspended a wire above it, just dipping into the mercury. Faraday found that when he passed an electric current through the wire it would spin around the magnet. He called this "electro-magnetic rotations," and it was to form the basis of the electric motor. In fact, a changing magnetic field generates an electric field and vice versa.

Faraday could not find time to continue his work on electromagnetism immediately, and it fell to the American scientist Joseph Henry (1797–1878) to develop the first powerful electromagnet in 1825. He found that by winding insulated wire around a magnet and running a current through the wire he could greatly enhance the power of

the magnet. He built one electromagnet that could lift a weight of 3,500 pounds (nearly 1600 kilograms). Henry also went on to lay the foundations for the electric telegraph. He strung a 1 mile (1.7 km) length of fine wire through the Albany Academy and then passed electricity through the wire, successfully using it to power a bell at the other end. Although it was Samuel Morse (1791–1872) who went on to develop telegraphy, Henry had proved that the concept was sound.

FIELDS AND FORCES

A field is the way in which a force is transmitted across distance. A magnetic field is that area in which a magnetic force operates. It is usually shown in the form of lines radiating from the north pole of a magnet to its south pole. The strength of an electromagnetic or gravitational force reduces relative to the square of the distance from the source—so at twice the distance from the source, the force has only a quarter of its original strength. The inverse square law relating to forces was first noted by Newton in relation to gravitational force.

If one name stands out in relation to electricity it is probably that of Michael Faraday. Although he was too busy to continue work on electromagnetism in the 1820s after his first experiment, he returned to the topic in 1831 and discovered the principle of electric induction. Faraday wound two coils of wire around opposite sides of an iron ring and then passed a current through one wire. This magnetized the ring and briefly induced a current in the other wire coil, making the first electric transformer. Six weeks later Faraday invented the dynamo, in which a permanent magnet is pushed backward and forward through a wire coil, inducing a current in the wire. Faraday's law of induction states that magnetic flux varying in time produces a proportional electromotive force. All electricity generation is based on this principle. Faraday also introduced the terms electrode, anode, cathode, and ion, speculating that part of a molecule was involved in moving electricity between the cathode and the anode. The true nature of ionic solutions and their conductivity was finally explained by Arrhenius, who was awarded the Nobel Prize for this work in 1903.

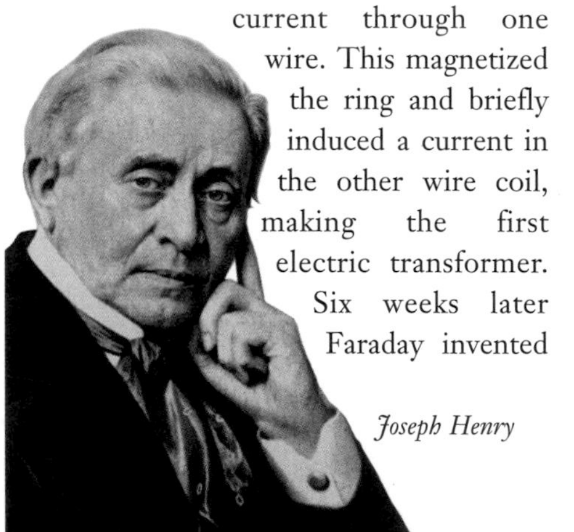
Joseph Henry

Dawn of a New Electromagnetic Age

Building on the practical work of Ørsted and Faraday, James Clerk Maxwell brought mathematics to bear on the relationship between electricity and magnetism. The result was four earth-changing equations, published in 1873, that demonstrated electromagnetism is a single force. Einstein

"This was the first discovery of the fact that a galvanic current could be transmitted to a great distance with so little a diminution of force as to produce mechanical effects, and of the means by which the transmission could be accomplished. I saw that the electric telegraph was now practicable... I had not in mind any particular form of telegraph, but referred only to the general fact that it was now demonstrated that a galvanic current could be transmitted to great distances, with sufficient power to produce mechanical effects adequate to the desired object."

Joseph Henry

considered Maxwell's equations to be the greatest discovery in physics since Newton had formulated the law of gravity. Electromagnetism is now recognized as one of the four fundamental forces that keep the universe in order—the others being gravity, and the strong and weak nuclear forces that operate within and between atoms. On the smallest scale, electromagnetic forces bind ions together into molecules and provide the attraction between the electrons and nucleus of an atom.

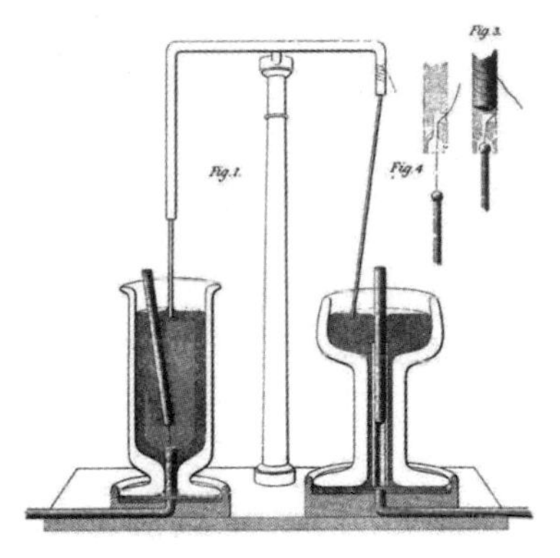

Faraday's equipment demonstrated electromagnetic rotation.

Maxwell explained how both electric and magnetic fields arise from the same electromagnetic waves. A varying electric field is accompanied by a similarly varying magnetic field that lies at right angles to it. He found, too, that the wave of undulating electromagnetic fields travels through empty space at the speed of light—approximately 186,000 miles (300,000 km) per second. This was a stunning discovery, and not everyone was happy with the conclusion that light is part of the electromagnetic spectrum. Einstein incorporated Maxwell's work into his theories of relativity, saying that whether a field was electric or magnetic depended on the viewer's frame of reference. Looked at from one frame of reference the field is magnetic. Looked at from a different frame of reference, the field is electric.

A magnetic field is demonstrated by the arrangement of compass needles around a magnet.

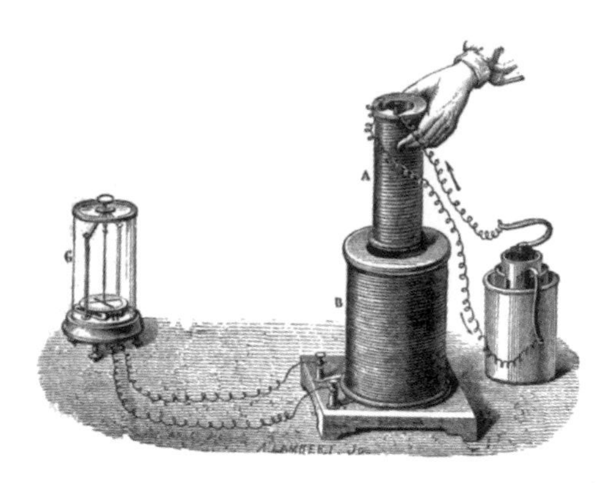

Faraday's apparatus showing electromagnetic induction between two coils of wire. A liquid battery on the right provides a current and the small coil is moved in and out of the large coil by hand to induce a current in the large coil, indicated by the galvanometer on the left.

FARADAY'S LAWS OF ELECTROMAGNETIC INDUCTION

1. An electromagnetic field is induced in a conductor when the magnetic field surrounding it changes.
2. The magnitude of the electromagnetic field is proportional to the rate of change of the magnetic field.
3. The sense of the induced electromagnetic field depends on the direction of the rate of change of the magnetic field.

More Waves

Although Maxwell predicted the existence of radio waves, they were not observed until the German physicist Heinrich Rudolf Hertz (1857–94) generated electromagnetic waves with a wavelength of 13 ft (4 m) in his laboratory in 1888. Hertz did not recognize the significance of radio waves and, when asked about the impact his discovery would have, said, "Nothing, I guess." As well as generating radio waves, Hertz discovered that they could be transmitted through some materials but bounced off others—a quality that would later lead to the development of radar. The discovery of radio waves made Maxwell's explanation of electromagnetic radiation irresistible. Over the coming years, the discovery of microwaves, X-rays, infrared, ultraviolet, and gamma rays completed the electromagnetic spectrum.

The next form of energy to be discovered was X-rays. Although the German physicist Wilhelm Conrad Röntgen (1845–1923) named and described X-rays and is usually credited with their discovery in 1895, he was not in fact the first to observe them. They were first detected in around 1875 by his fellow countryman the physicist Johann Wilhelm Hittorf (1824–1914). Hittorf was one of the inventors of the Crookes tube, an experimental device used to investigate cathode rays. It consists of a vacuum within which a stream of electrons flows between a cathode and anode and is a precursor of the cathode ray tube used in televisions before the advent of the modern plasma screen. Hittorf found that when he left photographic plates near the Crookes tube some were later found to be marked by

"Convert magnetism into electricity!"

Michael Faraday's to-do list, 1822; accomplished in 1831

MICHAEL FARADAY (1791–1867)

Born in London to a poor family, Faraday left school at 14 and was apprenticed to a bookbinder, educating himself by reading the science books he worked on. After hearing four lectures given by Humphry Davy at the Royal Institution in 1812, Faraday wrote to Davy asking for a job. Davy refused the request initially, but the following year employed him as a chemical assistant at the Royal Institution. At first Faraday just helped other scientists, but then he began conducting his own experiments, including those with electricity. In 1826 he started the Royal Institution's Christmas lectures and Friday Evening Discourses—both still continue today. Faraday gave many lectures himself, becoming established as the leading scientific lecturer of his day. He discovered electromagnetic induction in 1831, laying the foundation for the practical use of electricity, which had previously been considered an interesting phenomenon, but of little real use.

In recognition of his achievements, Faraday was twice offered the presidency of the Royal Society (and twice turned it down), and was offered a knighthood (which he also rejected). He ended his days in Hampton Court Palace in a home that was the gift of Queen Victoria's consort, Prince Albert.

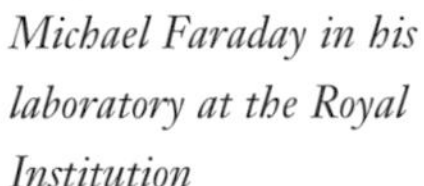

Michael Faraday in his laboratory at the Royal Institution

shadows, but he didn't investigate the cause. Other scientists skirted around X-rays, too, before Röntgen produced his famous X-ray photograph of his wife's hand and explained the phenomenon. Röntgen had his laboratory notes burned after his death so it's impossible to know exactly what happened, but it seems that he was investigating cathode rays using a screen painted with barium platinocyanide and a Crookes tube wrapped in black. He spotted a faint green glow from the screen and realized that rays of some kind were passing through the cardboard from the tube and making the screen glow. He investigated the rays, and published his findings two months later.

MAXWELL'S EQUATIONS

$$\oint \mathbf{E} \cdot d\mathbf{A} = \frac{q_{enc}}{\varepsilon_0}$$

$$\nabla \cdot \mathbf{E} = \rho/\varepsilon_0$$

Maxwell's first equation is Gauss's law, which describes the shape and strength of an electric field, showing that it reduces with distance following the same inverse square law as gravity.

$$\oint \mathbf{B} \cdot d\mathbf{A} = 0$$

$$\nabla \cdot \mathbf{B} = 0$$

The second equation describes the shape and strength of a magnetic field: the lines of force always go in loops from the north to the south pole of a magnet (and a magnet must always have two poles).

$$\oint \mathbf{E} \cdot d\mathbf{s} = -\frac{d\Phi_{\mathbf{B}}}{dt}$$

$$\nabla \times \mathbf{E} = -\frac{\partial \mathbf{B}}{\partial \mathbf{t}}$$

The third equation describes how changing electric currents create magnetic fields.

$$\oint \mathbf{B} \cdot d\mathbf{s} = \propto_0 \varepsilon_0 \frac{d\Phi_{\mathbf{E}}}{dt} + \propto_0 i_{enc}$$

$$\nabla \times \mathbf{B} = \propto_0 \varepsilon_0 \frac{\partial \mathbf{E}}{\partial \tau} + \propto_0 j_c$$

The fourth equation describes how changing magnetic fields create electric currents, and is also known as Faraday's law of induction.

Radiation

When the French physicist Henri Becquerel (1852–1908) heard about X-rays, in 1896, and that they came from a bright spot on the wall of a Crookes tube, he suspected phosphorescent objects may also emit X-rays. Becquerel was professor of physics at the French Museum of Natural History and so had access to a large collection of phosphorescent materials. He discovered that, if allowed to absorb energy from sunlight for a while, these materials would glow in the dark until the energy was used up. He then found that if he wrapped a photographic plate in dark paper to exclude light and placed it over a dish of phosphorescent salts that had been "charged" in the sun, light areas appeared on the plate. By putting a metal object between the plate and the dish, he produced a shadow image of the object on the photographic plate, just like Röntgen's X-ray plates. In a later experiment, he prepared his set-up and planned to leave it in the sun. Although Paris had no sunshine for several days, Becquerel decided to develop the plate anyway, expecting to find nothing. To his astonishment, he found an image—the uranium salts he was using seemed to emit X-rays without exposure to the sun, apparently violating the law of the conservation of energy to produce energy

> *"We can scarcely avoid the conclusion that light consists in the transverse undulations of the same medium which is the cause of electric and magnetic phenomena."*
>
> James Clerk Maxwell, *ca.*1862

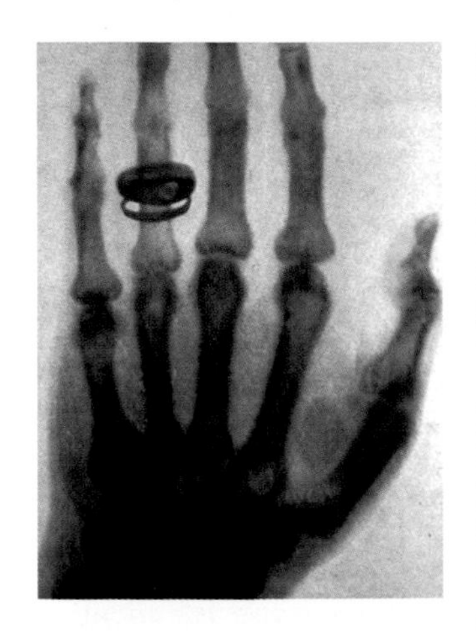

Röntgen's X-ray of his wife's hand was the first X-ray picture produced; her wedding ring is clearly visible.

"It's of no use whatsoever [...] this is just an experiment that proves Maestro Maxwell was right—we just have these mysterious electromagnetic waves that we cannot see with the naked eye. But they are there."

Heinrich Hertz, on his discovery of radio waves in 1888

LET THERE BE LIGHT

The first public electricity supply was in Godalming, Surrey, England where electric street lighting was installed in 1881. A waterwheel on the River Wey drove a Siemens alternator that powered arc lamps in the town, providing electricity to several shops and other premises.

from nothing. He investigated further and found the radiation was not the same as X-rays as it could be deflected by a magnetic field and so must consist of charged particles. He did no more work on the subject, though, leaving the field clear for the Polish-born experimental physicist Marie Curie.

Marie Curie (1867–1934) was working on her PhD on "uranium rays" when she discovered that the ore from which uranium was derived, pitchblende, is more

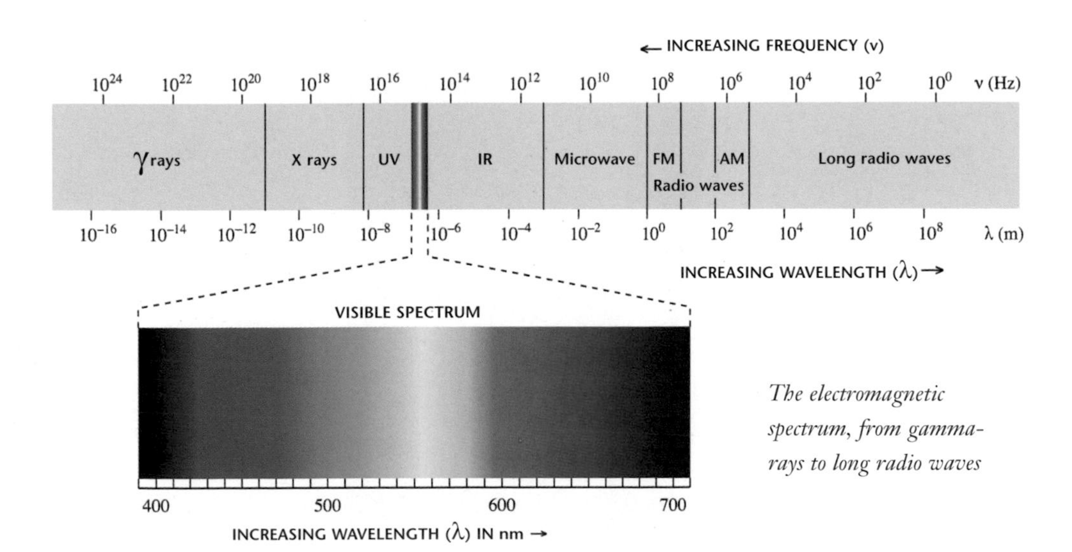

The electromagnetic spectrum, from gamma-rays to long radio waves

BECQUERELS FOREVER

The professorial chair in physics at the French Museum of Natural History was essentially hereditary. The chair was founded for Antoine Becquerel (1788–1878) in 1838 and was occupied by a Becquerel without interruption until 1948, when the incumbent failed to produce a son to pass it on to.

radioactive than the element itself. This suggested that there were other, more radioactive elements in the ore. With her husband Pierre, she extracted two such elements—polonium and radium. It took four years from her discovery in 1898 to extract a tenth of a gram of radium, using tons of pitchblende. Pierre discovered that a gram (0.035 ounces) of radium could heat one and a third grams of water from freezing point to boiling point in an hour—and could keep on doing it, again and again. It looked like energy for nothing, an astonishing discovery.

The Curies did not know what form of energy radioactivity actually was. That discovery fell to the New Zealand-born British chemist and physicist Ernest Rutherford (1871–1937), working at the Cavendish Laboratory in Cambridge. Rutherford was the first person to be admitted to Cambridge as a research student, rather than progressing to research after having obtained a degree at the university. He turned up, on a scholarship from New Zealand, two months before Röntgen discovered X-rays, but only got the post by chance. He was one of two applicants for the scholarship and was not chosen, but the successful applicant then dropped out. Rutherford began working on radio waves, and may have achieved long-range transmission before Marconi, but as he wasn't interested in their commercial potential he didn't exploit his findings.

When Rutherford turned his attention to radiation, he found that the form Becquerel had discovered was made up of two different types: alpha radiation, which can be blocked by a sheet of paper or a few centimeters of empty air, and beta radiation, which can penetrate further into matter. In 1908, Rutherford showed that alpha radiation is a stream of alpha particles: helium atoms stripped of their electrons. Beta radiation consists of fast-moving electrons—like a cathode ray, but with more energy. In 1900 Rutherford discovered a third type of radiation, which he called gamma radiation. Like X-rays, gamma rays form part of the electromagnetic spectrum. They are high-energy waves with a wavelength even shorter than X-rays. Rutherford's work had taken him inside the atom, which is our next destination.

Ernest Rutherford

NEEDED—ATOMS

Work on thermodynamics in the late 19th century killed the caloric model of heat and led physicists such as the Austrian Ludwig Eduard Boltzmann and James Clerk Maxwell to believe that heat is a measure of the speed at which particles are moving, though they were not sure about the nature of the particles involved. The transfer of heat and conductivity of electricity could both only be fully understood once it became clear that they depended on an atomic model of matter. For electricity to travel through a conductor, electrons must pass between atoms; for heat to move from one place to another by conduction or convection, particles most move.

The acceptance of the atomic model of matter at the turn of the 20th century opened the door to exploring the inside of the atom, and that in turn led to a greater understanding of how energy behaves and is transmitted.

MARIE CURIE (MANYA SKLODOWSKA, 1867–1934)

Born in Warsaw, in Russian-occupied Poland, Manya Sklodowska had no chance of a university education in her homeland, so went to the Sorbonne in Paris to study. There she met and married Pierre Curie, who was already working on magnetic materials. Pregnancy delayed her PhD, which was on the topic of "uranium rays." She had to work in a drafty shed as academics feared the presence of a woman in the laboratory would lead to so much sexual tension that no work would get done, at least by the men. In 1898 she began work isolating the unknown radioactive elements in pitchblende (uranium ore). Her husband Pierre abandoned his own research to help her. They discovered two radioactive elements, polonium (named after Poland) and radium. In 1903, Marie and Pierre Curie were awarded the Nobel Prize for Physics, which they shared with Henri Becquerel. Only three years later, Pierre was killed after he slipped on a Paris street and his skull was crushed by the wheel of a passing horse-drawn carriage. He may have been suffering dizzy spells, a symptom of radiation sickness. Marie died of leukaemia in 1934, also a victim of radiation exposure. Her notebooks remain so radioactive that even today they must be stored in a lead-lined safe. She is the only woman to have received two Nobel Prizes (the second was for chemistry in 1911, also for her work on radioactivity).

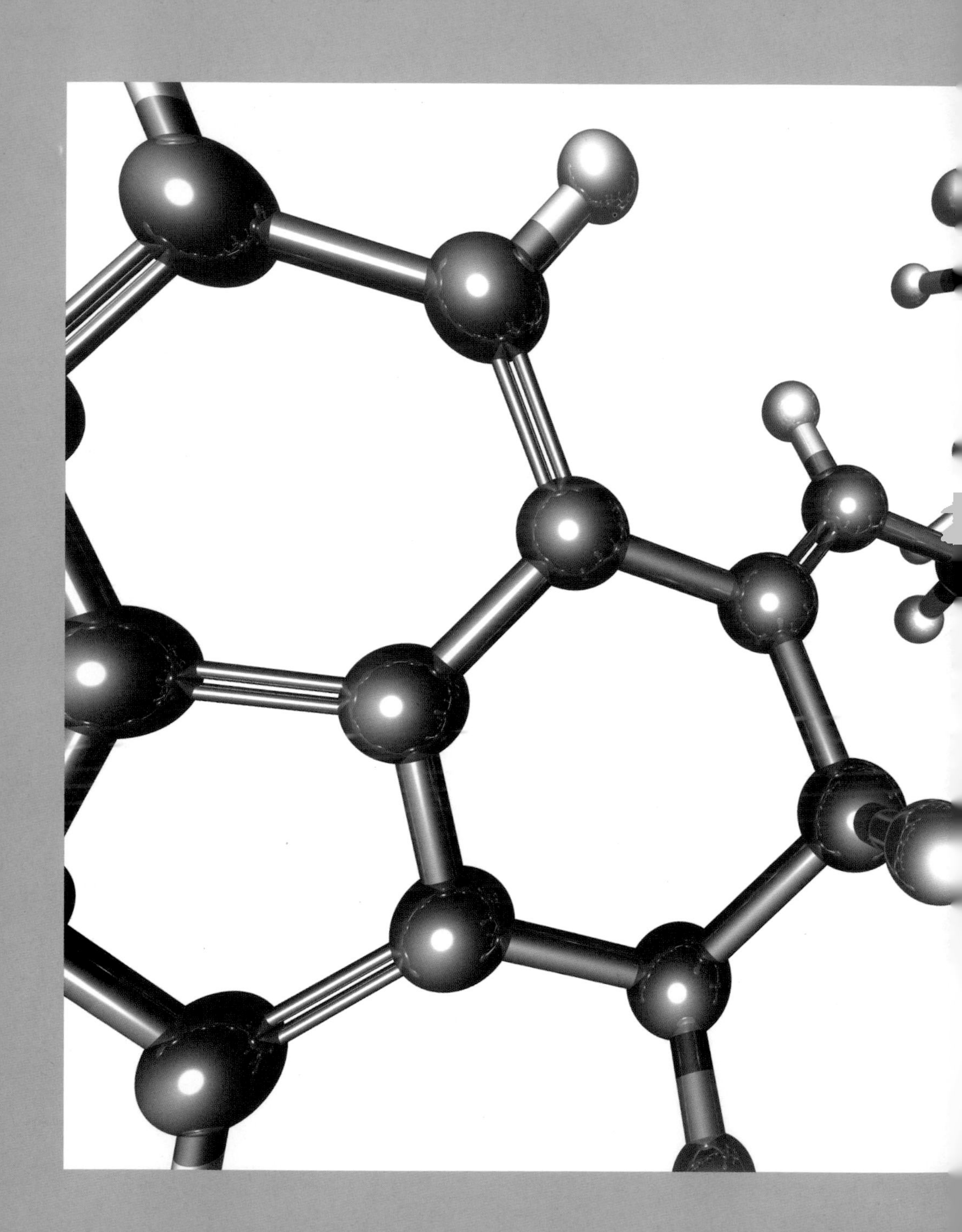

CHAPTER 5

Into the ATOM

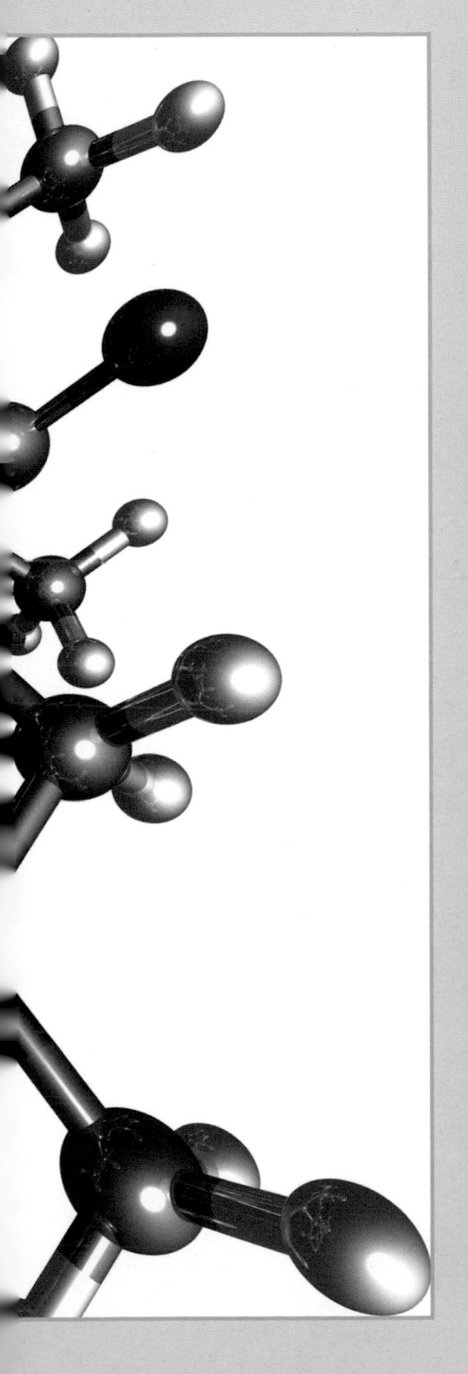

A belief in atoms as the building blocks of matter has an ancient history. Some Buddhist thinkers in the seventh century BCE believed all matter to be made of atoms, which they regarded as a form of energy. In Europe, pre-atomists such as Empedocles and Anaxagoras had also conceived of invisibly tiny particles of matter. These early philosopher-scientists arrived at their view through a process of deductive thought alone. Although atomism fell out of favor for many centuries it was, in the end, the model that would prevail, supported by experimentation and observation. But the early atomists weren't quite right. Their belief that atoms are the smallest, indivisible particles of matter proved to be incorrect, for atoms are made up of subatomic particles. As scientists probed inside the atom, it would prove to be a bizarre and unpredictable place.

The discovery of the atomic structure of matter opened the door to a whole new world for physicists.

Dissecting the Atom

John Dalton described his atomic theory in 1803, saying that elements are made of identical atoms that combine in ratios of whole numbers to form chemical compounds. The theory was not universally accepted until the French physicist Jean Perrin (1870–1942) measured the size of a water molecule more than a century later, in 1908, though many scientists did accept and work with the theory before this date. But even before the theory had been confirmed as fact, the premise that atoms cannot be subdivided was breaking down.

The British physicist Joseph John (J.J.) Thomson (1856–1940) discovered the electron in 1897 during his work on cathode rays and Crookes tubes (see page 104). Thomson found that cathode rays travel much more slowly than light and so cannot, as previously suspected, be part of the electromagnetic spectrum. He concluded that a cathode ray is actually a stream of electrons. The concept that the electron was a part of the atom that could break free and operate on its own overturned the previous belief that the atom was indivisible. In 1899, Thomson measured the charge on an electron and calculated its mass, reaching the startling conclusion that it is around one two-thousandth of the mass of a hydrogen atom.

Although Thomson was awarded the Nobel Prize for his work on the electron in 1906, its importance was not immediately obvious. In fact, physicists couldn't see the point of the electron and, indeed, the toast at the annual Cavendish Laboratory dinner in Cambridge was, "To the electron: may it never be of use to anybody."

Plum Puddings and Solar Systems

J.J. Thomson's model of the atom, proposed in 1904, has been called the "plum pudding" model because it resembles a ball of suet pudding studded with currants. He described the atom as a cloud of positive charge punctuated throughout with electrons. In a rather confusing example of recycling terminology, he called them "corpuscles." The positively charged part remained rather nebulous, while the electrons were well-defined currants stuck into it, possibly orbiting in fixed rings.

The plum-pudding model was disproved in 1909 by an experiment carried out by the German physicist Hans Geiger (1882–1945) and the New Zealander Ernest Marsden (1889–1970) at the University of Manchester, while working under the supervision of Ernest Rutherford. Their experiment involved directing a beam of alpha particles toward a very thin sheet of gold foil surrounded by a circular sheet of zinc sulfide. The zinc sulfide lit up when hit by alpha particles (helium nuclei). The experimenters expected to see the alpha particles passing through with little deflection and that the pattern they made after passing through the foil would give information about the distribution of charge within the gold

> *"The assumption of a state of matter more finely divided than the atom is a somewhat startling one."*
>
> J.J. Thomson

atoms. The results were a surprise. Very few particles were deflected at all, but those few were deflected through angles much greater than 90 degrees. Rutherford was expecting the experiment to endorse the plum-pudding model and was totally unprepared for this result. The only conclusion that he could draw was that the positive charge in the atom was concentrated in a tiny center, not distributed throughout the whole atom.

Rutherford was left with the task of coming up with a new model for the structure of the atom to replace the discredited "plum pudding." What he produced was a model with a tiny, dense nucleus surrounded by a lot of empty space, punctuated by orbiting electrons. He was not sure whether the nucleus had a positive or negative charge, but calculated its size to be less than 3.4×10^{-14} meters across (it's now known to be about a fifth of this). A gold atom was known to be around 1.5×10^{-10} meters in radius, making the nucleus less than 1/4000th of the diameter of the atom.

J.J.THOMSON (1856–1940)

Joseph John (J.J.) Thomson was the son of a bookbinder. He was too poor to take an apprenticeship as an engineer, and instead went to study mathematics on a scholarship at Trinity College, Cambridge, England. He would eventually become master of that college, establish the Cavendish Laboratory as the foremost physics laboratory in the world, and be awarded a Nobel Prize for physics for his work on the electron. Through experimentation with cathode rays, Thompson was able to identify the electron as a particle in 1897 and he then measured its mass and charge in 1899. In 1912 he showed how to use the positive rays that could be produced using a perforated anode in a discharge tube to separate atoms of different elements. This technique forms the basis of mass spectrometry, commonly used today to analyze the composition of a gas or other substance. Thomson was notoriously clumsy. Not only did he rely on his research assistants to handle delicate experiments for him, they even tried to keep him out of the laboratory in case he disrupted their equipment. But he was well liked and inspirational: seven of his research assistants and his own son went on to be awarded Nobel Prizes. Thompson was knighted in 1908.

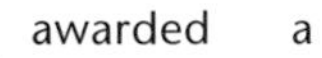

> *"The atoms of the elements consist of a number of negatively electrified corpuscles enclosed in a sphere of uniform positive electrification."*
>
> J.J. Thomson, 1904

THE SATURNIAN MODEL

In 1904 the Japanese physicist Hantaro Nagaoka proposed a model of the atom based on Saturn and its rings. This gave the atom a massive nucleus and orbiting electrons kept in place by an electromagnetic field. He came up with the idea after hearing Ludwig Boltzmann talking about the kinetic theory of gases and James Clerk Maxwell's work on the stability of Saturn's rings while he was touring Germany and Austria in 1892–96. Nagaoka abandoned the theory in 1908.

Rutherford hadn't finished with the atom. He proposed a structure in which the nucleus of the atom contained positively charged particles—protons, which he discovered in 1918—and some electrons. The rest of the electrons orbited the nucleus, he thought.

The Danish physicist Niels Bohr (1885–1962) refined Rutherford's model in 1913 in such a way that electrons were enabled to stay in orbit. He suggested that instead of wandering around the space outside the nucleus following any path they pleased, electrons are restricted to particular orbits and are physically incapable

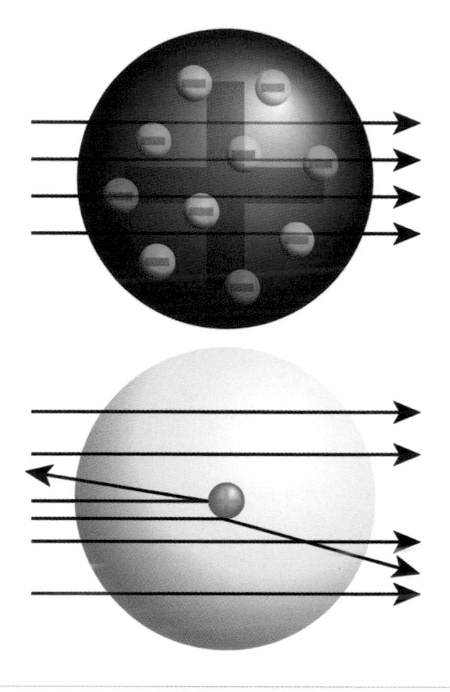

The top diagram shows the expected result of Rutherford's gold foil experiment, with alpha particles passing through the atom; the bottom diagram shows the surprising result—some particles were widely deflected.

> *"It was quite the most incredible event that has ever happened to me in my life. It was almost as incredible as if you fired a 15-inch shell at a piece of tissue paper and it came back and hit you. On consideration, I realized that this scattering backwards must be the result of a single collision, and when I made calculations I saw that it was impossible to get anything of that order of magnitude unless you took a system in which the greater part of the mass of the atom was concentrated in a minute nucleus. It was then that I had the idea of an atom with a minute massive centre, carrying a charge."*
>
> Ernest Rutherford

Saturn's rings provided a model for the atom for Nagaoka.

of constantly emitting radiation (which they would be able to do if the laws of classical physics applied). Bohr believed these orbits are circular and fixed, giving a planetary model of the atom, the electrons being the planets orbiting a nucleus that is the equivalent of the sun. Unlike planets, though, electrons can jump between orbits, releasing or absorbing a specific quantity (a quantum) of energy each time, according to whether they are moving toward or away from the nucleus.

According to Bohr's model, the single electron of the hydrogen atom, for example, can exist in only a limited number of orbits. Each orbit represents a particular energy level. The lowest level is called the ground state and is the closest point the electron gets to the nucleus. When the hydrogen atom absorbs a photon of light the electron jumps to an orbit of larger radius (higher energy level). Which orbit or level it jumps to depends on the energy contained in the photon. When the atom emits that photon, the electron jumps back to its previous orbit (lower energy level).

Each orbit, he argued, had enough room for a certain number of electrons only, so they couldn't all crowd as close to the nucleus as possible, however much they may like to. This means that the orbits fill up from the inside outward.

The electron absorbs or releases a single photon or quantum of energy only when it makes a "quantum jump" between orbits. The amount of energy—or wavelength—of the energy absorbed or released is determined by the

Niels Bohr in 1935

orbit. It looked like a neat bit of trickery—but when Bohr tested his theory he found that hydrogen atoms emit energy at the wavelength his maths predicted if the electrons could jump between their prescribed orbits, which he called shells. Furthermore, Bohr's model explained why hydrogen—and indeed every element—produces a unique absorption and emission spectrum. This principle lies at the heart of spectroscopy, used by astronomers to reveal the chemical composition of stars.

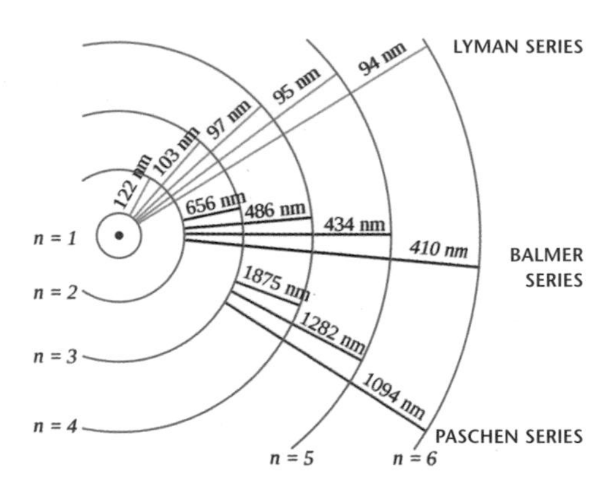

The electron shell transitions of hydrogen, with their associated energies

Quantum Solace

When Max Planck talked of quanta as a way of moving energy in little parcels, he didn't really mean anyone to take the quantum seriously; it was a theoretical solution that he assumed would be replaced as soon as someone worked out the mathematics that explained what was really happening. But he had hit on something that had apparently turned out to be true, however unlikely it seemed. And it was not only true, but the basis of a whole new type of physics that operates in the bizarre world of subatomic particles. Quantum mechanics—which accounts for the behavior of particles on a tiny scale as Newtonian mechanics accounts for the behavior of larger systems—began with Planck's quick-fix of quanta. It's a topsy-turvy realm of seeming impossibilities and mind-boggling suggestions.

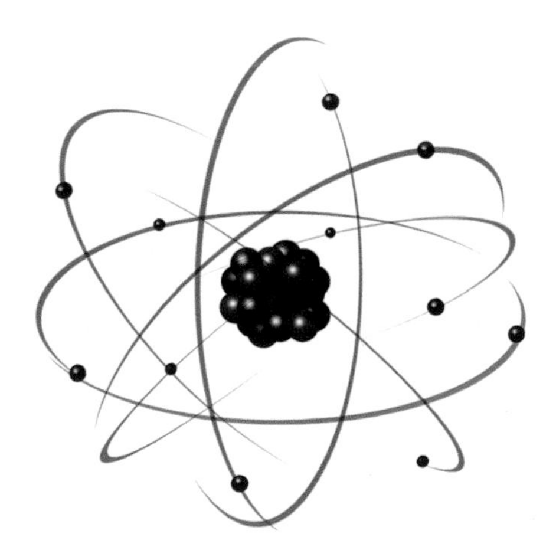

In Bohr's model of the atom, electrons generally stay firmly in their allotted shells and orbit the nucleus.

Einstein took quanta seriously. His work on the photoelectric effect (see page 53) drew on Planck's use of quanta, but applied it to light. Einstein suggested that a photon could have enough energy to knock an electron out of an atom; a stream of knocked-out electrons produced an electric current. His idea was unpopular at first, as it flew in the face of Maxwell's equations and the received wisdom that light is a wave. Here, for the first time, physics came up against wave-particle duality—something that sometimes acted like a wave and sometimes like a particle.

Solar panels make use of the photoelectric effect to generate electricity from photons striking a semiconductor.

Smart Light

Even more intriguing was the discovery that light seems to "know" how to behave to please experimenters. When an experiment is designed to test the behavior of light as a wave, light acts like a wave. When an experiment tests the behavior of light as a particle, light behaves as a particle. It's not possible to catch it out. If a beam of light shines through two slits on to a screen, the

NIELS BOHR (1885–1962)

The work of Danish physicist and philosopher Niels Bohr was key to the development of quantum mechanics, turning a sketchy hypothesis into a workable concept. Through quantum physics he expanded Rutherford's theory of atomic structure and explained the spectrum of hydrogen. But he never underestimated the complexities involved, once remarking that, "You never understand quantum physics, you just get used to it." Bohr began his studies at Copenhagen University before moving to England to work at Cambridge and Manchester. He then returned to Copenhagen to found the Institute of Theoretical Physics. In 1922 he was awarded the Nobel Prize for Physics. During the Second World War he joined the team developing the atom bomb. His career might have taken a very different path—in 1908 he narrowly missed selection as goalkeeper in Denmark's national football side. Soccer's loss was physics' gain.

Albert Einstein (left) with Niels Bohr

standard interference pattern is produced, with dark and light stripes. As the light is dimmed further and further, there comes a point when individual photons pass one at a time to the screen, making a flash each time one appears. However, collectively, the picture that is built up is still the interference pattern. The photons seem to "know" whether one or two slits are open and if two slits are open the interference fringe still builds up, however slowly photons are fired at the screen. Each individual photon seems able to go through both slits simultaneously. If one slit is closed, even after a photon has started its journey, photons only go through the open slit. Taking it even further, if there is a detector on one of the slits to discover whether the photon has gone through that slit or the other, the photons, as if reluctant to be caught out, stop producing interference patterns—suddenly they act as particles.

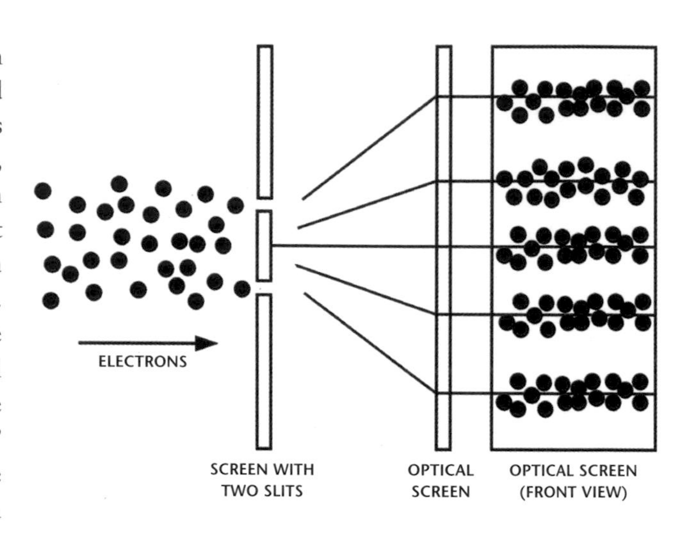

The double-slit experiment that produces diffraction patterns with light can also be produced by scattering electrons, showing that they, too, can behave as waves.

As though that weren't strange enough, in 1924 French physicist Louis-Victor de Broglie (1892–1957) suggested that the particles that make up matter could also behave as waves. This would mean that wave-particle duality is everywhere and all matter has a wavelength. In 1927, his odd idea was supported by electrons apparently behaving as waves and diffracting the way light does. Since then, larger particles—protons and neutrons—have also been seen to act like waves.

De Broglie's work was his PhD thesis. In it he suggested that electrons were waves running around the orbits they were allowed to occupy, and the energy levels of the permissible orbits were harmonics of the wave so that the waves always reinforced each other. The theory could be tested, he said, by showing that electrons are diffracted by a crystal lattice. This was successfully demonstrated in 1927 by two separate experiments, one in the USA and one in Scotland. De Broglie and two of the three men who carried out the experiments were awarded the Nobel Prize for Physics in 1937 for the work.

The importance of de Broglie's work was that it showed that wave-particle duality applies to all matter. His equation states that the momentum of a particle (of anything), multiplied by its wavelength, is equal to Planck's constant. Because Planck's constant is very small, the wavelength of anything larger than a molecule is small compared to

its actual size. We wouldn't worry about the wavelength of a bus or a tiger, for instance. As we consider smaller and smaller particles, their wave properties become more important.

> **GIANTS AND THEIR SHOULDERS**
> Classical physics began in earnest with Newton, and his "miraculous year" (*annus mirabilis*) of 1666. The rebirth of physics that kick-started quantum mechanics began with the publication of Albert Einstein's special theory of relativity in 1905. Both scientists were building on the work of many earlier scientists who had made these moments of revelation possible. Their discoveries reverberated through the years that followed.

ANOTHER NEWTONIAN MOMENT

That particles might in fact behave like waves does not seem so impossible after Einstein explained, in 1905, why.

In an appendix to his special theory of relativity, Einstein included an early (less succinct) form of this equation, which translates into real words as

Energy = mass × the speed of light squared
which is now more familiar as
$E=mc^2$

It is a world-shifting result, as important as Newton's *Principia.* Einstein's equation was saying that energy *is the same* as matter but in a different form. Matter can be converted into a very large quantity of energy. This lies at the heart of nuclear power and nuclear weapons, both of which trade in the energy that can be released by messing with the nuclei of atoms.

There was a fundamental problem with

> **WAVES AND PARTICLES**
> Wave-particle duality is mirrored perfectly in the history of Nobel Prizes for physics. One of the men who shared the Nobel Prize with de Broglie [pictured right] for demonstrating the wave properties of electrons was George Thomson. He was the son of J.J. Thomson who had been awarded the Nobel Prize in 1906 for demonstrating that electrons were particles. Neither is considered to have been wrong; both explanations are still accepted. (Nobel Laureates aren't allowed to be wrong.)

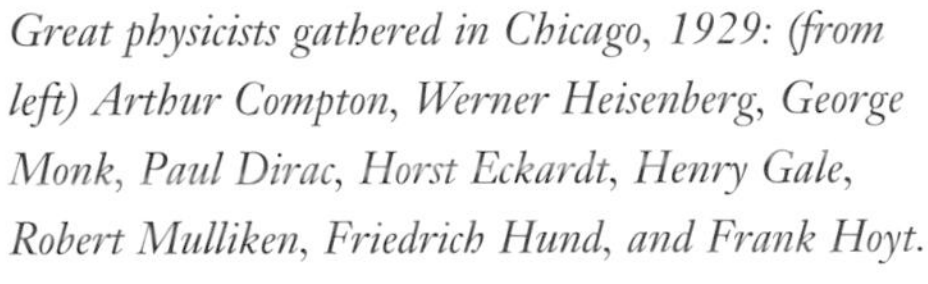
Great physicists gathered in Chicago, 1929: (from left) Arthur Compton, Werner Heisenberg, George Monk, Paul Dirac, Horst Eckardt, Henry Gale, Robert Mulliken, Friedrich Hund, and Frank Hoyt.

Schrödinger's model gives precise results without the limitations inherent in the Bohr model. However, replacing certainty with probability threw quantum physicists into turmoil.

At the same time that Schrödinger was pursuing the electron-as-wave model, the German physicist Werner Heisenberg (1901–1976) was making his own mathematical model of the electron, but favoring its particle-like properties as it made quantum leaps between orbitals. He, like Schrödinger, published in 1926. The British physicist Paul Dirac (1902–84) developed a third, more mathematical and theoretical model at the same time.

Werner Heisenberg, left, swimming with friends. Even nuclear physicists relax sometimes.

In fact, Dirac went on to show that the other two models—Heisenberg's and Schrödinger's—were actually equivalent, and that all three of them were saying the same thing in slightly different ways. The three men won a Nobel Prize for their contributions to quantum mechanics.

Can We Be Certain?

Heisenberg's uncertainty principle, stated in 1927, asserts that we cannot know everything about a particle. He saw that one consequence of quantum mechanics is that it is impossible to measure all aspects of a particle at the same time. If we measure its position and speed, we can know both within certain limits, but increasing the accuracy of one measurement makes the other less certain. The very act of observing its position makes its speed less certain. This is a fundamental property of the quantum description of measurement and can't be avoided by changing the method or tools of observation.

Heisenberg originally argued the uncertainty principle using a thought experiment. For example, we could measure the position of a moving particle by shining a light on to it, in which case there will be one of two outcomes. A photon of light may be absorbed, causing an electron in the atom to jump to another energy level, in which case we have altered the atom and our measurement is false. Alternatively, a photon is not absorbed but passes straight through, in which case we have made no measurement at all.

The uncertainty principle is further complicated if we try to treat both the "particle" and the photon as wave-particles.

WHERE CAN I GO FROM HERE? THE ELECTRON'S CONUNDRUM

The whole of quantum mechanics can be constructed starting from the uncertainty principle. Recalling the original problem with the atomic model presented by Newtonian mechanics, of why the electrons don't just fall into the nucleus and have done with it, Heisenberg's principle gives an explanation. The momentum of a particle in a particular orbit is known, so its position can't be precisely known—it is just somewhere in the orbit. However, if the particle fell into the nucleus its position would be known—but so would its momentum, as that would be zero. By falling into the nucleus, the electron would violate the uncertainty principle. It simply isn't allowed to do so. In fact, the smallest orbit in an atom (look at the orbit of the electron in a hydrogen atom) is as small as it possibly can be without violating the uncertainty principle—the maths works. The size of atoms and indeed their very existence are determined by the uncertainty principle.

Heisenberg realized that the uncertainty principle affected not just the present, but also the past and future. Because a position was always and is only a collection of probabilities, pinning down the path of a particle is not what it appears. As Heisenberg said, "The path comes into existence only when we observe it." The future path similarly can't be predicted with certainty.

Newtonian physics deals with certainties, with cause and effect, a

> *"Anyone who is not shocked by quantum theory has not understood it."*
>
> Niels Bohr

deterministic model in which knowledge allows prediction. The new quantum mechanics appeared to overturn all that, at least at the atomic level. It was far from popular in some circles—even Einstein distrusted it, saying "God doesn't throw dice," though he had to accept the mathematics. Indeed, since the start of the 20th century, the use of mathematical models has steadily taken over from experimental physics that could be tested in the laboratory. The thought experiment, supported by mathematical calculations, had become the mainstay of the new, largely theoretical physics.

The Copenhagen Interpretation

While Schrödinger tended to focus on the wave aspects of the wave-particle duality, Heisenberg concentrated more on the particle. Heisenberg presented his work in the form of matrices, while Schrödinger worked with probability theory. Around these two approaches, two separate camps of physicists emerged, each thinking that the other approach was wrong.

In 1927, Bohr, Heisenberg, and the German-born physicist Max Born (1882–1970) worked together to produce a synthesis of the apparently contradictory aspects of quantum theory, known as the "Copenhagen Interpretation." This says that it is not that atomic particles or photons "choose" whether to act as a wave or particle at any point, or that they are actually one or the other: instead, the features that make them appear to act like one or the other are two sides of the same coin. Which one we see, and how we interpret their behavior, depends on what we are looking for and how we are observing them. Light exists as both a wave and a particle simultaneously but only appears as one or the other when we measure it. The act of measuring or observing determines the outcome because of the type of observation we choose to make. At the point when the measurement is made and the wave-ness or particle-ness is determined, the wave function is said to collapse. More precisely, it instantly and discontinuously changes to the wave function that would be associated with the result of the measurement.

Bohr recognized the importance of the uncertainty principle, but went further than Heisenberg in pointing out that it is not a problem that comes from the physical interference involved in measurement, but a more fundamental issue—the very act of making a measurement changes the situation (or system) being examined. This casts doubt on the whole premise of the scientific method. There can be no objective observer if the act of measurement or observation itself affects the outcome.

A Cat in a Box

Bohr's explanation did not please everyone. Schrödinger showed his disdain by describing a thought experiment to demonstrate the absurdity of the Copenhagen Interpretation. In Schrödinger's experiment, a cat is shut in a box with a device that consists of a tiny amount of radioactive substance, a Geiger

Schrödinger's cat, both dead and alive, in a box both with and without poison

counter, a small flask of hydrocyanic acid, and a hammer. The equipment is set up so that if an atom of the radioactive substance decays, the detection of the particle released will cause the hammer to break the flask and the cat will be poisoned by the gas. There is an equal probability that an atom will or will not decay, and the cat cannot interfere with the equipment. The cat is left in the box for an hour. By the end of the hour the chances are 50:50 that it is alive (or dead). Following Bohr's argument and the Copenhagen Interpretation, the state—dead or alive—of the cat is not fixed until we look in the box. This, he said, was ridiculous.

Many Universes

Another response to the unsavory idea that everything exists in a probability cloud until observed was the "many worlds" model proposed in 1957 by the American physicist Hugh Everett III (1930–82). This suggests that there is an infinite number of parallel universes that account for all possible outcomes to all possible questions. At decision-making (or observing) points, a new universe splits off. If nothing else, this helps us come to grips with infinity. If a new universe splits off each time you choose between tea and coffee, or a tadpole swims to the left or right, or a falling branch does or does not fall on a roof, there must be a great many universes—somewhere.

Quantum Entanglement: The Einstein-Podolsky-Rosen Paradox

Albert Einstein was one of those who would not accept the Copenhagen Interpretation. In 1935, Einstein and the American physicists Boris Podolsky (1896–1996) and Nathan Rosen (1909–95) concocted the so-called EPR paradox. Suppose a

Erwin Schrödinger

stationary particle decays, producing two other particles. They must have equal and opposite angular momentum so that these cancel each other out (conservation of angular momentum) and all their other quantum properties must similarly balance to conserve the properties of the parent particle. This link between the particles must continue to exist after they have been emitted and gone their separate ways. If we measure a property for one particle, we collapse the wave function for the same property in the other particle—it is affected instantly and inevitably.

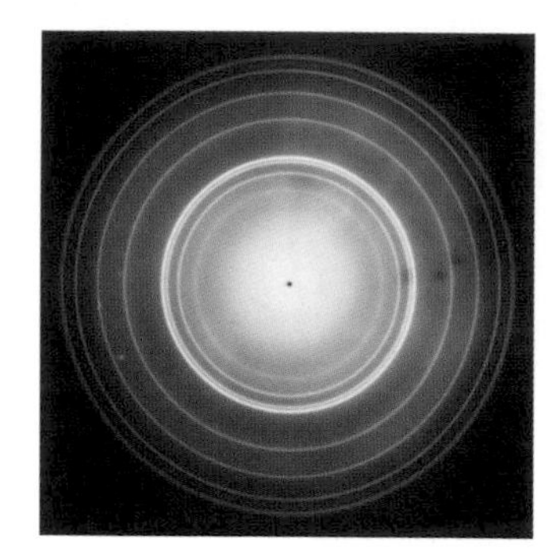

The electron diffraction pattern of beryllium

Just like Schrödinger's cat, Einstein's entangled particles were deliberately devised to show the absurdity of the Copenhagen Interpretation but ended up strengthening it.

The entanglement of particles has since been demonstrated to exist, with particles separated by many kilometers. Entanglement

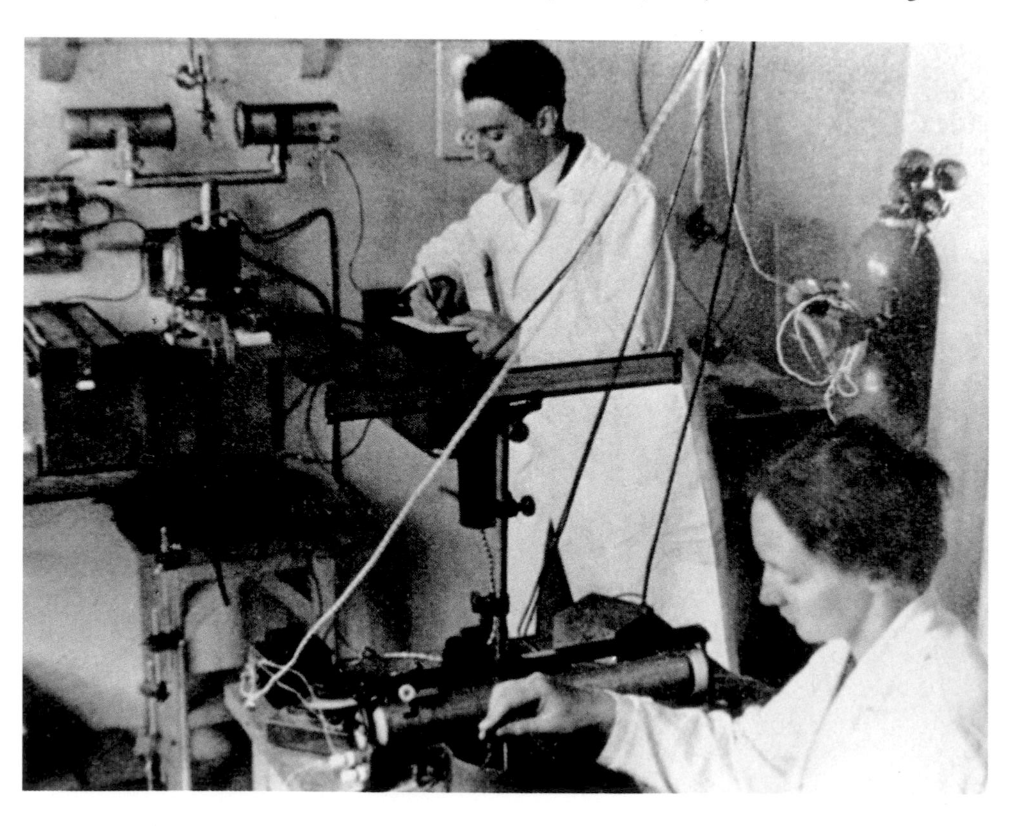

Frederic Joliot and Irène Joliot-Curie working in their laboratory

James Chadwick earned a Nobel Prize for his work on the neutron carried out in February 1932.

may even be put to practical use, offering new fast methods of computing (using "qubits," or quantum bits), instantaneous communication, and encryption. In fact, entanglement offers a way of transmitting information faster than the speed of light.

The Search for More Atomic Particles

It has long been known that electrons could be knocked out of the atom fairly easily as that is how they were discovered in 1897. In the early 1930s, Walter Bothe (1891–1957) and Irène Joliot-Curie (1897–1956, daughter of Marie and Pierre Curie) and her husband Frédéric (1900–1958) discovered that alpha-particle radiation fired at beryllium produced another type of radiation. This type was good at knocking something out of other elements, but it wasn't immediately clear what. The Joliot-Curies announced their results in January 1932. The English physicist James Chadwick (1891–1974) repeated the experiments immediately and explained the effect by suggesting that the alpha particles were knocking bits out of the nucleus of the beryllium atoms. At first, he had thought these "bits" were proton-electron pairs, as they had no (or a balanced) electrical charge.

Throughout the 1920s Chadwick had been looking for a neutral particle, which he had expected to take the form of a proton and electron bound together. But his most important work, for which he was awarded the Nobel Prize in 1935, was in the end crammed into a few hectic days in February

CANDIDATES FOR NEUTRONHOOD

Two years before Chadwick claimed the name "neutron" for his uncharged particle in the nucleus, the Austrian physicist Wolfgang Pauli (1900–1958) had used the same name for a theoretical particle that he suggested was emitted from the nucleus during beta-radiation. His idea had so little impact at the time that Chadwick was able to poach the name with no problem. The existence of Pauli's particle was eventually confirmed in the 1950s and is now called the neutrino (see page 135).

> *"We might in these processes obtain very much more energy than the proton supplied, but on the average we could not expect to obtain energy in this way. It was a very poor and inefficient way of producing energy, and anyone who looked for a source of power in the transformation of the atoms was talking moonshine. But the subject was scientifically interesting because it gave insight into the atoms."*
>
> *The Times,* 12 September 1933, on a speech by Ernest Rutherford on atomic energy

working together, developed a model of radioactive decay in 1903. They explained that an atom of a heavy element could be unstable and decay by losing an alpha particle (helium nucleus) or having a neutron decay into a proton and emit a beta particle (electron). In both cases the number of protons in the nucleus changes, so the atom becomes a different element. They predicted that the decay of radium would produce helium, a result that Soddy achieved in 1903 while working with the Scottish chemist Sir William Ramsay (1852–1916) in London. In 1913, Soddy stated that emitting an alpha particle reduced the atomic number by two (as two protons are lost), whereas emitting a beta particle increased it by one (as a neutron decays into an electron, which is lost, and a proton, which remains, so increasing the atomic number). Soddy came up with the name "isotopes" to describe variants of an element with different atomic masses.

Enrico Fermi

In 1919, Rutherford found that if he bombarded nitrogen with alpha particles, it turned into an isotope of oxygen, losing a hydrogen nucleus (a single proton) in the process. This was the first artificial transmutation of one element into another—a goal held dear by alchemists throughout the centuries, though with the more ambitious target of changing base metal into gold. Rather than the first step into a new world of alchemy, it was the first step into the realm of nuclear physics.

Between 1920 and 1924, Rutherford and Chadwick demonstrated that most of the lighter elements will emit protons if blasted with alpha particles.

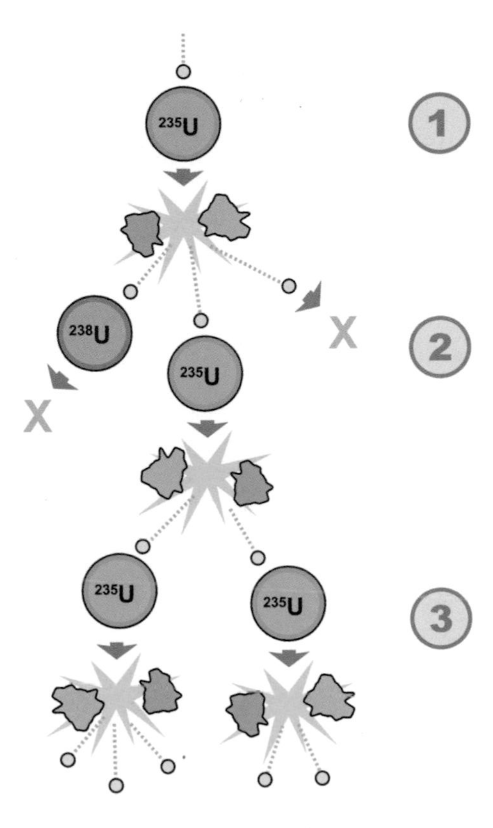

The chain reaction produced by the decay of uranium-235 induced by bombardment with neutrons

Irène and Frédéric Joliot-Curie discovered induced radioactivity in 1934, finding that by blasting some elements with alpha-particles they could turn them into unstable radioactive isotopes that would then decay. The Italian physicist Enrico Fermi (1901–1954) extended their research, using slow neutrons to produce more effective induced radioactivity. Blasting uranium with neutrons, Fermi thought he had created a new element, which he called hesperium. In 1938, though, a group of four German and Austrian scientists discovered that in fact Fermi's technique had split the uranium nucleus into two roughly equal parts. This process is nuclear fission.

The Hungarian physicist Leó Szilárd (1898–1964) realized that the neutrons released by a nuclear fission reaction could be used to spark the same reaction in other atoms, leading to a self-sustaining chain reaction. Szilárd was in London when he was incensed by an article in *The Times* dismissing the possibility raised by Rutherford that the energy within atoms could be harnessed for practical use. While

Harnessing the Chain Reaction

The transformation of one element into another can be triggered artificially and may be the source of immense power. The energy released in the detonation of an atomic bomb, or harnessed in a nuclear power station, comes from a nuclear chain reaction with the particles emitted by one decaying atom used to trigger another.

The world's first nuclear reactor becomes self-sustaining in Chicago, 1942 (no photographers were present).

"THE WORLD SET FREE"—OR NOT

Leó Szilárd was inspired by a novel written by the British writer H.G. Wells, called *The World Set Free* (1914), in which a new type of weapon, an "atomic bomb," wrought devastation. Wells' fictional atomic bombs continued to explode over a period of days. This led Szilárd to begin serious consideration of harnessing nuclear chain reactions to make a real atomic bomb. Szilárd moved to the USA in 1938 and a year later persuaded Albert Einstein to join him in writing to US President Franklin D. Roosevelt, urging his government to set up a research program to develop an atomic bomb, to counter the risk of Nazi Germany developing nuclear weapons first. It became the Manhattan Project. Szilárd envisaged the project as a way of protecting the world against the destruction described by Wells as he hoped the bomb would be held as a threat and not actually used. He became increasingly distressed as control of the research passed to the military and urged a show test of an atomic bomb to demonstrate its power to the Japanese in order to secure surrender without loss of life, a suggestion the US government rejected. Atom bombs were dropped on the Japanese cities of Hiroshima and Nagasaki in 1945, causing great devastation and many thousands of deaths. After the war, Szilárd predicted the nuclear stalemate that would characterize the Cold War. He turned away from physics to concentrate on research into molecular biology.

"We turned the switch and saw the flashes. We watched them for a little while and then we switched everything off and went home... That night, there was very little doubt in my mind that the world was headed for grief."

Leó Szilárd, on successfully starting a chain reaction using uranium in Columbia University, Manhattan, in 1938

walking to work at St Bartholomew's Hospital and waiting for traffic lights to change on Southampton Row in Bloomsbury, Szilárd worked out how a nuclear chain reaction might work. He filed a patent for it the following year. Indeed, Szilárd originally held the patents for both the chain reaction and the nuclear reactor (with Enrico Fermi), though he surrendered the patent for nuclear chain reactions to the British Admiralty in 1936. Szilárd was a prime mover in the development of the atom bomb (see box).

Frédéric Joliot-Curie produced experimental evidence for the chain reaction in 1939 and scientists in many countries (including the United States, the United Kingdom, France, Germany, and the Soviet Union) clamored for money to research nuclear fission. The first nuclear reactor to go live was Chicago-Pile-1 in December 1942, built to produce plutonium for use in nuclear weapons.

The End of the Classical Atom

Already, with Bohr's model, it was impossible to account for the behavior of the atom in terms of classical physics. The

Atom bombs were detonated over Hiroshima (left) and Nagasaki (right) in August 1945.

tiny nucleus contains the protons and neutrons, held together by the strong nuclear force. In the rest of the empty space, electrons whizz around in their designated shells, never wandering from their orbits but able to jump from one to another in the right circumstances. What the ancients, with their concept of the indivisible atom, would have found hard to grasp was not only that the atom comprised electrons, protons, and neutrons, but that the protons and neutrons in their turn can be broken down still further. The second half of the 20th century saw the discovery of quarks, held together by a force mediated by gluons. Intriguingly, this is the strong force—the same force responsible for binding protons and neutrons together. Indeed, the binding of protons and neutrons is something of a residual effect.

The strong force acting on quarks is altogether more interesting. Rather than diminishing over distance, the force becomes stronger until it reaches a maximum that it exerts over all distances substantially larger than the size of a proton or neutron. Gluons were first detected in 1979 using the electron-positron collider PETRA in Germany.

Protons and neutrons are examples of hadrons, all of which are made up of either three quarks (baryons) or one quark and one antiquark (mesons). Experiments at the Stanford Linear Accelerator Center in 1968 revealed that the proton is not indivisible but comprises smaller point-like objects that Richard Feynman called "partons."

The quark model had been proposed in 1964, but partons were not immediately identified with quarks. Quarks come in six flavors: "up," "down," "top," "bottom," "strange," and "charm" ("top" and "bottom" are sometimes called "truth" and "beauty"). Antimatter quarks—antiquarks have antiflavors, which gives rise to such odd concepts as the "antistrange" quark and the "anti-up" quark. In normal life these might be called "mundane" and "down," but in the odd world of quarks, "down" is not the same as "anti-up."

Both protons and neutrons are baryons and are the only stable hadrons, though neutrons are only stable inside the nucleus of an atom. There are around 40 known or predicted types of baryon and around 50 known or predicted types of meson. They have bizarre names, such as "double charged bottom Omega" (a baryon of unknown mass or duration). Some are very short-lived (if they exist at all)—such as the delta baryon, which lasts only 5.58×10^{-24} seconds (so it would take around 30 times as many delta particles as there are stars in the universe to last for a single second.) The first mesons to be discovered were kaons and pions, found in cosmic rays in 1947.

The large number of subatomic particles is beyond the scope of the current book, but suffice to say that there are many as yet undiscovered or proven, some with unknown properties and functions.

Matter and Antimatter

In 1927, Paul Dirac published a definitive wave equation of the electron that fully accommodated the requirements of the special theory of relativity (see page 122). Surprisingly, though, it had two solutions; one described the familiar electron and the other something equivalent to the electron

THE QUARK OF A DUCK

The name "quark" was chosen by one of the two people who proposed their existence independently in 1964, Murray Gell-Mann. He named it after the sound made by ducks, wanting it pronounced "kwork," but could not immediately decide on a spelling. He settled on "quark" after finding the word in James Joyce's *Finnegans Wake:*

Three quarks for Muster Mark!
Sure he has not got much of a bark
And sure any he has it's all beside the mark.

but with a positive charge. At first, Dirac tried to fit it to the proton, but that had too much mass. Other investigations suggested that by using enough energy, a pair of particles could be created with opposing electric charge but identical mass. In 1932 and 1933, Carl Anderson found traces of a positively charged particle as predicted by Dirac. He called it a positron. Others recognized it as the first antimatter particle to be discovered. The positron has since found a practical application in a medical imaging technique called PET (positron emission tomography) scanning. We now know that all particles have matching anti-matter particles with exactly opposite properties.

Murray Gell-Mann gave quarks their name.

Ghost Particles

One of the more intriguing and elusive particles is the neutrino, first suggested by Wolfgang Pauli in 1930. He needed it to balance an equation. When the nucleus of a radioactive atom decays, the energy released should be equal to that originally present. But Pauli found this was not the case. More energy was being lost than could be measured, which meant something was being emitted that was not being picked up by detectors. Pauli was aware that during beta decay the electrons emitted could apparently have any amount of energy up to a maximum for each particular type of nucleus. But if this were really the case, it would violate the law of the conservation of energy. Pauli's radical solution was to suggest the existence of another, uncharged particle that was not quantized and could carry any amount of kinetic energy up to a preset maximum. He called his potential particle a neutron, though two years later Chadwick would take this name for the particle we now know as a neutron.

In 1933, Enrico Fermi came up with the name "neutrino" for Pauli's mystery particle. Fermi suggested that a neutron decays into a proton and an electron (which it also does if it is taken outside the atomic nucleus) and *also* a new type of uncharged particle, the neutrino. The neutrino was then emitted along with an electron during beta decay.

Neutrinos remained theoretical until the American physicists Frederick Reines (1918–98) and Clyde Cowan (1919–74) detected them in 1953. They used large water tanks near a nuclear reactor as "neutrino collectors." They calculated the

> *"I have done something very bad today by proposing a particle that cannot be detected. It is something no theorist should ever do."*
>
> Wolfgang Pauli, diary, 1930

reactor would be generating 10 trillion neutrinos per second and managed to trace three an hour. Clearly a lot were escaping, but the few they found provided the much-needed proof that neutrinos really do exist.

Neutrinos have negligible mass and no charge so they pass through everything they meet without hindrance. Indeed, if a beam of neutrinos were fired at a wall of lead 3,000 light years thick, half would get through without being stopped. There are neutrinos left over from the Big Bang, neutrinos emitted by the sun, and streaming from exploding stars. Around 100 trillion neutrinos pass through your body every second. Atoms are mostly empty space so there is plenty of room for the neutrinos to zip through everything, and because they have no charge they are not deflected or distracted by electrons or protons.

About 10 years after the first neutrino was discovered, a specialized detector was installed in a gold mine in South Dakota. The detector comprised a vast tank filled with dry-cleaning fluid rich in chlorine. When a neutrino collides with a chlorine atom, it creates radioactive argon. Every few months, a search of the tank would reveal about 15 argon atoms, showing that 15 neutrinos had collided with chlorine atoms over that time. The detector was used continuously for more than 30 years.

Today there are many more neutrino detectors built deep underground, some in old mines, others under the ocean and even beneath the Antarctic ice. It is no trouble to the neutrinos to get to the detectors, but the shielding prevents scientists confusing them with cosmic rays (larger particles that

The MINOS (Main Injector Neutrino Oscillation Search) detector in the Soudan Underground Mine State Park, used for investigating neutrinos

are stopped by the intervening matter). The Super-K detector in Japan uses 50,000 metric tons of water in a domed tank with 13,000 light sensors. The sensors detect a blue flash whenever a neutrino collides with an atom in the water and creates an electron. By tracing the exact path the electron takes through the water, physicists can work out the direction the neutrino came from and hence deduce its source. Most come from the sun. In 2001, physicists discovered that neutrinos come in three "flavors." There are many more types than they had realized, but they have only been discovering those that create electrons when they interact with matter. The discovery of flavors has a further implication—it means neutrinos have mass. A detector to measure the mass of a neutrino will soon go into operation in Germany.

A ROUNDABOUT ROUTE

The Karlsruhe Tritium Neutrino Experiment (KATRIN), which will be used to calculate the mass of a neutrino, was built 250 miles (400 km) from Karlsruhe, Germany, where it will operate. However, it was too large to transport through the narrow roads, so was carried by boat along the River Danube, into the Black Sea, through the Mediterranean, around Spain, through the English Channel, and into the Rhine, to Leopoldshafen, Germany where it then continued by road. The journey took two months and covered 5,600 miles (over 9,000 km).

Feynmann's work on the spin and rotation of electrons was prompted by seeing a spinning plate and pondering the "wobble" as he watched the pattern.

"While I was eating lunch, some kid threw up a plate in the cafeteria. There was a blue medallion on the plate, the Cornell sign, and as he threw up the plate and it came down, the blue thing went around and it seemed to me that the blue thing went around faster than the wobble, and I wondered what the relation was between the two. I was just playing, no importance at all, but I played around with the equations of motion of rotating things, and I found out that if the wobble is small the blue thing goes around twice as fast as the wobble goes round.

"I started to play with this rotation, and the rotation led me to a similar problem of the rotation of the spin of an electron according to Dirac's equation, and that just led me back into quantum electrodynamics, which was the problem I had been working on. I kept continuing now to play with it in the relaxed fashion I had originally done and it was just like taking the cork out of a bottle—everything just poured out, and in very short order I worked the things out for which I later won the Nobel Prize."

The Last Lost Particle

Antimatter and neutrinos were theorized before they were found. The hunt is on now for another theorized particle, the Higgs boson. Sometimes called the "God particle," the Higgs boson is the last particle in the so-called Standard Model of the physical world that has yet to be found. The Higgs boson does not have to exist in all models of physics, and in some models

RICHARD FEYNMANN (1918–88)

Born in New York, Feynmann was introduced to science early by his father, who made uniforms for a career but was interested in science and logic. Feynmann studied at Massachusetts Institute of Technology (MIT) and Princeton before working on the Manhattan Project to develop the atom bomb during the Second World War. He later joined the California Institute of Technology. Feynmann was a charismatic and popular lecturer with a range of unusual hobbies and interests, including playing bongo drums in a bar. He developed the mathematical theory of particle physics and demonstrated that the interaction between electrons (or positrons) can be considered in terms of the electrons exchanging virtual photons and showed these interactions in the form of "Feynmann diagrams." Famously, he had a van decorated with Feynmann diagrams, which is still in existence in a garage in California. He also pioneered quantum computing and came up with the concept of nanotechnology. Niels Bohr sought out Feynmann for one-to-one discussions of physics because everyone else was so in awe of Bohr that they would not contradict him or point out flaws in his arguments.

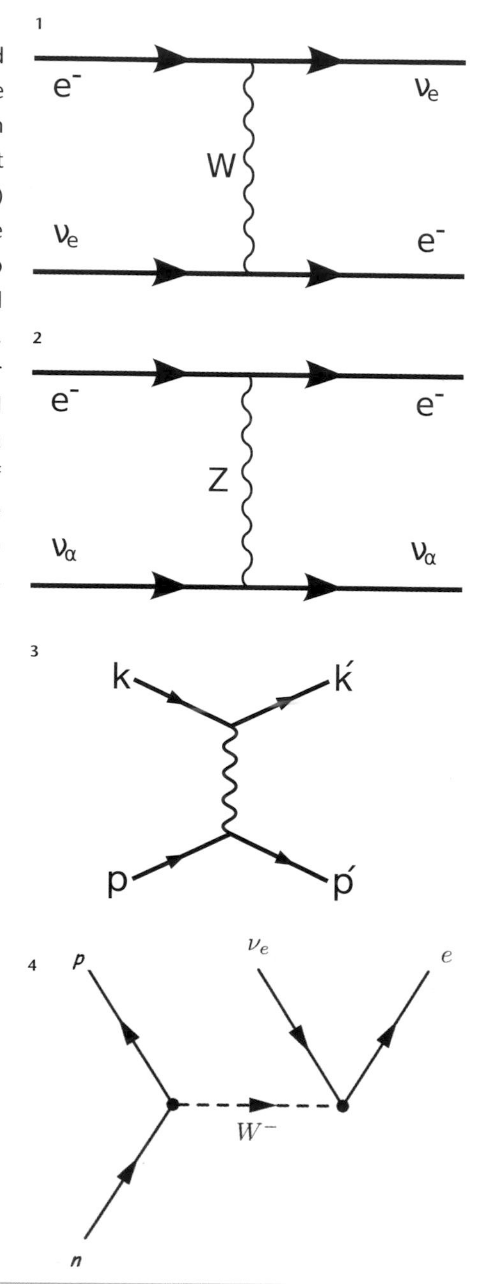

Feynmann diagrams of:
(1) neutrino interaction with matter with charged current; (2) neutrino interaction with matter with neutral current; (3) a scattering process; and
(4) neutron decay

Tunnel of the LHC at CERN

there may be more than one type of Higgs boson. Finding out whether or not the particle exists will help scientists to decide which of the suggested models is most likely to be correct. The Higgs boson is thought to be a component of the Higgs field. Passing through the Higgs field confers mass on particles. If the Higgs boson exists, it is an integral part of matter and is present everywhere. The first full description of the particle was given by Peter Higgs in 1966.

The search for the Higgs boson requires the use of large-scale colliders, such as the Large Hadron Collider (LHC) at Cern in Switzerland and Tevatron at Fermilab in the United States. There are several ways in which a hadron collider may produce a Higgs boson by smashing protons together at high velocities.

Particles from the Stars

Large hadron colliders try to emulate conditions that existed near the very start of the universe, with particles forced together under immense pressure. The fact that we have any idea at all of what may have happened near the start of the universe is the result of thousands of years observing and theorizing about the stars and space, an activity that no doubt began before recorded history with our earliest ancestors gazing in wonder at the sky and making up stories to explain what they saw.

RENAMING THE NONEXISTENT

Many scientists object to the popular term "God particle" for the Higgs boson. The most popular suggestion in a competition to rename it in 2009 was the "champagne bottle boson," but other contenders included the "mastodon," the "mysteron," and the "nonexiston."

CHAPTER 6

Reaching for the STARS

It's impossible to know when humans first looked up at the stars and wondered about them. Some were inspired to see pictures—constellations—in the pattern of the 4,000 or so stars visible to the naked eye, and then it must have been a small step to invent stories to accompany the pictures. Some of these stories became the basis of religious beliefs and attempted to explain the inexplicable—the origins of the world, the reason for the seasons, the movement of the stars and planets across the sky. Other people, it seems, were inspired to look for more rational explanations. They observed, counted, measured, and eventually made predictions. They doubtless tested and refined their predictions as passing time threw up problems with their models. These early astronomers were the first scientists. They were not in conflict with the religious traditions of their cultures, but worked hand in hand with them, predicting the movement of the heavenly bodies to produce calendars with religious as well as practical applications.

The Milky Way is our home in the universe, but only one of hundreds of billions of galaxies.

There are 3,000 prehistoric standing stones at Carnac in France.

Stars and Stones

Some of the oldest human structures may show evidence of careful observation of the movement of moon, stars, and planets across the sky. The 3,000 stones of Carnac, in France, date from around 4500–3300 BCE, and may have had astronomical significance. The circle of standing stones at Stonehenge in southern England, erected 3000–2200 BCE, might have served as a celestial observatory: the midsummer sun rises in approximate alignment with the central axis of Stonehenge. Earth's precession (the way our planet wobbles on its axis as it spins) means that Stonehenge would have been less accurately aligned 4,000 years ago than it is today, but it may still have provided astronomical data good enough for farming purposes and worship. Other investigators have found much more precise alignments with different celestial movements, including the moon and the planets, and have suggested that Stonehenge represents the result of decades or even centuries of astronomical observation.

The Great Pyramids at Giza in Egypt are more precisely aligned. Completed around 2680 BCE, all four sides of the three pyramids are astronomically oriented north–south and east–west to within a small fraction of a degree. The positions of the pyramids may mirror the central stars in the constellation Orion, with other pyramids possibly corresponding to other stars in Orion and the Nile corresponding with the Milky Way. The earliest certain depiction of astronomy from Ancient Egypt is the ceiling of the tomb of Senenmut, chief architect and astronomer during the reign of Queen Hatchepsut (*ca.*1473–1458 BCE). Several constructions built by the Mayan people in South America line up with the Pleiades star cluster and Eta Draconis (a star in the Draco constellation).

Early Stargazers

There are no contemporary records to support an astronomical use or correlation

Stonehenge near Salisbury, England, may have had astronomical uses in prehistoric times.

The Great Pyramids at Giza, Egypt, appear to be aligned with the stars and points of the compass.

for Stonehenge and the pyramids, but the earliest astronomers who did leave records date from around the same period. Chinese astronomers began to observe the sky using specially built observatories from around 2300 BCE. The first account of a comet was recorded in 2296 BCE, of a meteor shower in 2133 BCE, and of a solar eclipse in 2136 BCE. Chinese astronomy served astrology, with star-gazers needing to predict eclipses and other celestial phenomena in order to pick out propitious times to stage royal events and battles, and predict the future success and health of the Emperor. Failure could be fatal—at least two astronomers are known to have been beheaded in 2300 BCE for making inaccurate predictions of a solar eclipse. A tomb in Xishuipo, Henan province, dating from around 6,000 years ago, contained clamshells and bones forming the images of three constellations from Chinese astronomy, the Azure Dragon, the White Tiger, and the Northern Dipper. Oracle bones 3,200 years old bearing the names of stars relating to the 28 lunar mansions also survive. The Chinese believed that alignments in the heavens indicated or predicted significant events on Earth. From the 16th century BCE until the end of the 19th century CE,. nearly every dynasty appointed officials to observe and record astronomical events and changes, leaving an invaluable record for today's historians of astronomy.

The fertile crescent of Mesopotamia (now Iraq) was home to several early civilizations, beginning with the Sumerians around 2600 BCE. Tens of thousands of Sumerian clay tablets dating back to 2400 BCE include the earliest farmers' almanac with astronomical data, explaining when to sow and harvest crops.

The Babylonians occupied the area in about 1600 BCE. Their astronomers had state support for activities such as working out calendars and making astrological predictions. They compiled star catalogs, and began to keep long-term records of planetary motions and of solar and lunar eclipses that helped them to make approximate predictions of eclipses. They seem to have discovered the 223-month cycle of lunar eclipses. By 800 BCE they had fixed the locations of Venus, Jupiter, and Mars in relation to the stars and recorded the apparent retrograde (backward) motion of the planets.

The Babylonians developed a 12-month calendar with a bonus 13th month added occasionally to keep the years consistent. In some parts of Babylon, there was also a seven-day week. The Babylonians also divided the circle into 360 degrees, and from this derived a division of the day into 12 "kaspu," during which the sun spanned

Chinese constellations the Azure Dragon and White Tiger depicted on tiles

30 degrees of the sky. They used the arc of one degree as a unit to measure angular space.

Having a system for measuring angles enabled the Babylonian astronomers to measure the retrograde movement of the planets. From records kept on clay tablets over centuries, they could predict planetary positions and retrograde motions, even without understanding how or why the movements took place. They made no attempt at scientific explanations or models, as their predictions served practical and religious purposes only.

From Watching to Thinking

While the Chinese, Sumerian, and Babylonian astronomers were rigorous in recording the stars and events, the Ancient Greeks took a more theoretical and scientific approach, attempting to explain and model the behavior of the celestial bodies.

Around 500 BCE, Pythagoras suggested that the world is a globe, rather than flat, and in the fifth century BCE Anaxagoras proposed that the sun is a very hot rock, and that the moon is a chunk of the Earth. In 270 BCE Aristarchus said that Earth revolves around the sun. Previously, people believed the Earth to be the center around which the moon, sun, planets, and stars revolved. Aristarchus made the first calculation of the size of the sun and moon and their distance from Earth, and concluded that since the sun is much larger than Earth it is relatively unlikely that that the sun is the subordinate body, in orbit around the Earth.

Working from the length of time it took for a lunar eclipse to take place, Aristarchus

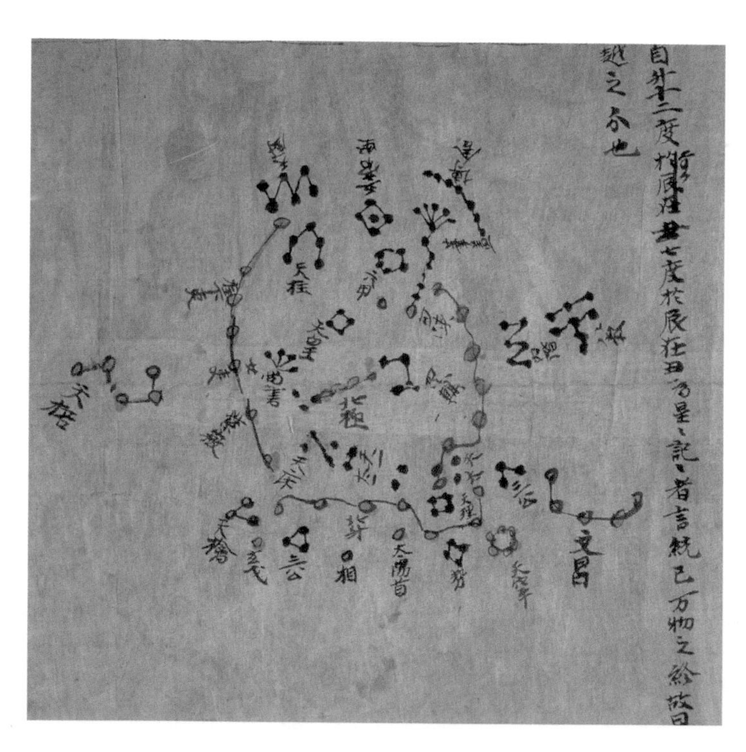

The Chinese Dunhuang star map was created in CE700.

> *"Then whether it would seem more probable, that the aequator of the terrestrial globe, in a single second (that is, in about the time in which one walking quickly will be able to advance only a single pace) can accomplish a quarter of a British mile (of which sixty equal one degree of a great circle on the Earth), or that the aequator of the primum mobile in the same time should traverse five thousand miles with celerity ineffable... swifter than the wings of lightning, if indeed they maintain the truth which especially assail the motion of the earth."*
>
> Edward Wright, in the introduction to William Gilbert's *De magnete* (1600) explaining why it is more likely that the earth turns on its axis than the sun revolves around the earth every 24 hours

calculated the distance from the Earth to the moon at about 60 times the radius of Earth, which fits the modern figure. He decided that the sun is 19 times as far from Earth as the moon is, and about 10 times the diameter of Earth, although he wasn't as accurate with these figures. Sadly, Aristarchus' conclusions were not accepted by his contemporaries. One argument was that if Earth moved around the sun, it would sometimes be much farther from the stars and their size would appear to vary. In fact, of course, Earth is so far from the stars that the distance the Earth travels is tiny by comparison and makes no difference to the apparent size of the stars, but otherwise it was quite a sensible point. But such distances were inconceivable at the time and Aristarchus' model was rejected. It would be 1,800 years before it would come back into favor.

Hipparchus invented the armillary sphere, shown here with the astronomer.

Hipparchus—The Greatest Astronomer of Antiquity?

The Greek astronomer Hipparchus was born in Nicaea *ca.*190 BCE but spent most of his life in Rhodes. He has been called the greatest astronomer of antiquity, though very little of his work survives. He is known to us largely through

Chart showing the Ptolemaic universe, with the Earth at the center, 1660–1

Ptolemy's *Almagest*. He drew on the work of the Babylonian astronomers, forming a bridge between Babylonian and Greek scholarship in the field and apparently using some of their methods as well as their collected data.

Ptolemy with an armillary sphere

Hipparchus was a great observer of the heavens and is often credited with producing the first detailed star catalog. The Chinese work *The Gan and Shi Book of the Stars*, written during the fourth century BCE, records the positions of 121 stars. But Hipparchus noted the positions of 850 stars visible to the naked eye, classifying them into six groups according to their brightness. This system is still used today. He drew up a list of all the eclipses that had taken place over the previous 800 years, and noted a new star in the constellation of Scorpio in 134 BCE. He has also been credited with inventing trigonometry and perhaps the planispheric astrolabe. Ptolemy said that Hipparchus explained the circular motion of the sun and moon but that he did not have a model for the paths of the planets, although he organized the data about them and showed that they did not accord with contemporary theories. His most famous achievement is his discussion of how the points of solstice and equinox move slowly from east to west when set against the fixed stars—known as the precession of the equinoxes.

Hipparchus first measured the length of a year accurately, making it 365 days, 5 hours, and 55 minutes. He noticed that the seasons were of different lengths and calculated the length of a month so accurately that he was only one second out.

PTOLEMY'S SPHERES

It should have been Aristarchus' heliocentric model that came down to us from the ancient world, but its place was taken by one described by Ptolemy around 140 CE. It did not originate with him—he was presenting the consensus of contemporary opinion in his *Mathematical Compilation* (now known as the *Almagest* after a corruption of its Arabic title). According to Ptolemy, Earth sits at the center of a group of concentric spheres. On these spheres, the moon, sun, planets, and fixed stars rotate around the Earth. The Greeks believed the circle to be the perfect shape, and as the heavens were the realm of perfection, orbits must be circular. This did not account for the observed motion of the planets, though.

A LESS PLAUSIBLE MODEL

In Hindu mythology, the world is said to be supported in space by four elephants, who in turn are standing on the shell of a turtle. There is no known astronomical observation that supports this model. Terry Pratchett borrowed the Hindu legend in his Discworld novels. The answer to the obvious question of what the turtle stands on has often been given as "it's turtles all the way down," a response which has been attributed to many sources.

To make the model work, the circular orbits of the planets had to be shifted away from Earth. It was clear that Venus and Mercury were in orbit around the sun, so Ptolemy's model had them following a circular path around the sun, which was itself following a circular path around Earth. Mars, Jupiter, and Saturn—the other planets visible to the naked eye—were also given something else to orbit around, but it was not the sun. Ptolemy identified empty points that formed the focus of the orbits of these planets, and these empty points revolved around Earth following a circular path. This pattern of shifted circular orbits accounted, near enough, for the slightly wandering path of the planets, which seem sometimes to go backward (follow a retrograde path). The fixed stars were easier to account for—they were just splattered on to a distant sphere that revolved around Earth, providing a backdrop to everything else.

With increasingly accurate observations of the movement of the planets, it became clear that the Ptolemaic model did not fully account for their paths. More and more little fixes were added to tweak the model and keep it in agreement with observations, but eventually—after more than a thousand years—it had to be given up.

BRAHMAGUPTA (598–668 CE)

The Indian mathematician Brahmagupta was born in Bhinmal city in Rajasthan, Northwest India. He was the head of the astronomical observatory at Ujjain, and wrote four texts on mathematics and astronomy, one of which contains the first account of the number zero. Brahmagupta proposed that Earth rotates on its axis, demonstrated that the moon is not farther from the Earth than the sun, and argued that Earth is round rather than flat. To counter criticisms that if Earth were a globe everything would fall off, he described something akin to gravity (see quote, below). He gave methods for calculating the position of heavenly bodies and predicting eclipses. It was from Brahmagupta's works that the Arab astronomers learned about Indian astronomy. Kankah, who came from Ujjain in 770 CE at the invitation of the caliph al-Mansur, used the *Brahmasphutasiddhanta* of Brahmagupta to explain astronomy.

"All heavy things are attracted toward the center of the earth... The earth on all its sides is the same; all people on earth stand upright, and all heavy things fall down to the earth by a law of nature, for it is the nature of the earth to attract and to keep things, as it is the nature of water to flow, that of fire to burn, and that of wind to set in motion... The earth is the only low thing, and seeds always return to it, in whatever direction you may throw them away, and never rise upward from the earth."

Brahmagupta, *Brahmasphutasiddhanta,* 628 CE

INTO AND OUT OF DARKNESS

With the decline of the Hellenic world, astronomy went into its own period of eclipse. There are no great Roman astronomers, and little progress was made before the rise of Arab science and the foundation of the Baghdad school of astronomy in 813 CE by al-Ma'mun.

While nothing much was happening in Europe and North Africa, Indian astronomers were making and recording observations that would later feed into Arab astronomy. The very earliest Indian text on the stars, the *Vedanga Jyotisa*, dates from around 1200 BCE but it is an astrological rather than an astronomical work and its uses were principally religious. The *Aryabhatiya*, from 476–550 BCE, was the first true astronomical text to circulate in India. It had an influence on later Arab writers, and is the first to set the start of the day at midnight. It states that the world rotates on its axis, which is why the stars appear to move across the sky, and that the moon is lit by reflected light from the SUN.

ARAB ASTRONOMY

The Arab astronomers were the first to apply mathematics consistently to the movement of the stars and planets. Islamic astronomers were impelled by the need for a reliable calendar, for the need to pinpoint accurately the times for prayers at sunrise, midday, afternoon, sunset, and in the evening, and to be able to determine the direction of their holy city Mecca from any location. They looked to the heavens to help with these tasks, prompted by the words of the Qu'ran (Koran) to use the stars for navigation: "It is He who ordained the

Arabic celestial map for the northern hemisphere, 1275

stars for you that you may be guided thereby in the darkness of the land and the sea." The Qu'ran, too, encouraged confidence in empirical data and the evidence of the senses, whereas the Greek thinkers had placed greater emphasis on reason. The Qu'ran's injunction to observe, reason and contemplate led toward an approximation of the scientific method.

Islam is generally opposed to the use of astrology for the purposes of prediction. When an eclipse occurred during the death of Mohammed's son, he discouraged onlookers from drawing conclusions about God, saying, "An eclipse is a phenomenon of nature and has no relation to the birth or death of a human being." This set Arab astronomy apart from the Indian and Chinese traditions, both of which harnessed astronomy in the service of astrology and future-casting.

From approximately 700–825 CE, most Arab astronomers concentrated on assimilating and translating astronomical works from the Greeks, Indians, and pre-Islamic Persians (Sassanids). Their own new endeavors began at approximately the time the caliph al-Ma'mun established the House of Wisdom in Baghdad. The arrival of paper in Iraq from China during the eighth century, long before it arrived in Europe, made it especially easy to collect and disseminate knowledge, and from 825 CE until the sacking of Baghdad by the Mongols in 1258, the House of Wisdom was the intellectual center of the world.

The first major original Muslim work of astronomy was the *Zij al-Sindh* written by Muhammed ibn Musa al-Khwarizimi (*ca.*780–*ca.*850) in 830 CE. It consists of tables for the movements of the sun, the moon, and the five known planets. Al-Khwarizimi is remembered primarily as a mathematician (the Latinized form of his name, Algoritmi, gave us the term "algorithm"), and Arab advances in mathematics certainly aided the study of astronomy. He also improved the sundial

The demands of praying at the right time drove the Arabic development of the calendar, and so of astronomy.

The Crab nebula was created by a supernova event witnessed by astronomers in 1054.

and invented the quadrant, used to measure angles. Some time around 825–835 CE, Habash al-Hasib al-Marwazi produced *The Book of Bodies and Distances*, in which he gave improved estimates of some astronomical distances. He gave the moon's diameter as 1,887 miles (3,037 km) (it is actually 2,159 miles) and its distance from Earth as 215,208 miles (346,344 km) (it is 238,857 miles). In 964, the Persian astronomer Abd al-Rahman al-Sufi (903–86) recorded observations and drawings of the stars, giving their positions, magnitudes, brightness, and color. His book includes the first descriptions and pictures of the Andromeda galaxy. In 1006, the Egyptian astronomer Ali ibn Ridwan (988–1061) described the brightest supernova in recorded history, saying it was two to three times as large as Venus and a quarter as bright as the moon. It was also described by astronomers in China, Iraq, Japan, Switzerland, and perhaps by indigenous people in North America.

The advances that Arab astronomers could make were severely limited by their conviction that Earth was at the center of the celestial system and that infinity was impossible. However, Ja'far Muhammad ibn Musa ibn Shakir suggested in the ninth century CE that the heavenly bodies observe the same physical laws as operate on Earth (contrary to the belief of the ancients), and in the 11th century Ibn al-Haytham made the first attempt to apply the experimental method to astronomy. He used special apparatus to test how the moon reflected the light of the sun, varying the settings of his equipment and recording the effects. He suggested that the medium of the heavens is less dense than the air and refuted Aristotle's view that the Milky Way is a phenomenon of the upper atmosphere. By measuring its parallax, he deduced that it is very far from Earth. It was al-Biruni who discovered, in the same century, that the Milky Way is made up of stars. He also described gravity as "the attraction of all things toward the center of the Earth," and said that gravity exists within the heavenly bodies and celestial spheres (still working with the Ptolemaic model of the universe). Al-Haytham proposed that Earth rotates on its axis, an idea previously put forward by

EARLY TOOLS OF THE TRADE

The oldest known astronomical tools are Babylonian clay tablets showing three concentric circles divided into twelve sections. Each of these 36 fields shows the names of constellations and simple numbers, which may represent the months of the Babylonian calendar.

An astrolabe represents the positions of the planets and stars, based on the assumption that earth is at the center of the universe. Astrolabes were probably developed some time before the first century CE, although the earliest surviving instrument is Arabic and dates from 927–28 CE. Islamic tradition explains the origins of the astrolabe: Ptolemy was riding a donkey while looking at his celestial globe. He dropped the globe, and his donkey trod on it, squashing it flat and so giving Ptolemy the idea for the astrolabe.

An armillary sphere is a three-dimensional equivalent of an astrolabe, representing the planets and stars in a series of concentric rings with the earth at the center.

A quadrant is used to measure the elevation of a body above the horizon. The first recorded quadrant is mentioned by Ptolemy, around 150 CE. The Islamic astronomers built large quadrants, but the most famous was that used by Tycho Brahe (1546–1601) in his observatory at Uraniborg on the Danish island of Hven.

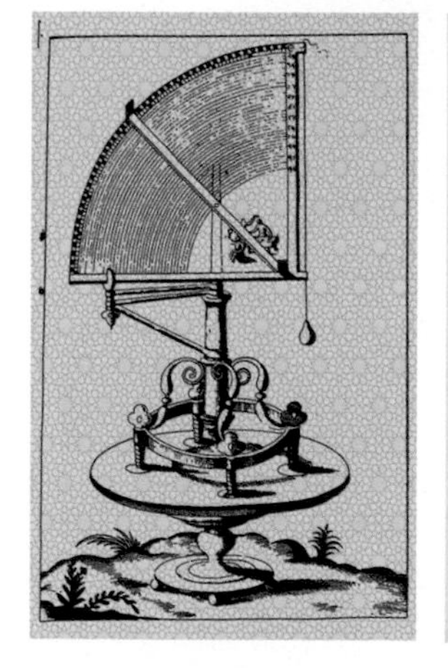

Ancient astronomical tools (clockwise from top): astrolabe, armillary sphere, quadrant

the Indian Brahmagupta. Al-Biruni found no mathematical problems with Earth's rotation when he commented on Brahmagupta's writings in 1030.

As with other aspects of Islamic science, rigorous investigation in astronomy was discouraged in Islam if regarded as trying to know the mind of God. Perhaps the most significant contribution of the Arab scholars from the eighth to the twelfth centuries was refinements in astronomical instruments and developments in mathematics. These paved the way for the European astronomers of the Renaissance who would rewrite the book of the heavens.

The Great Guest Star

For 23 days starting in July 1054, a star so bright that it could be seen in daylight blazed in the sky. Chinese astronomers referred to it as a "guest star" in the constellation of Taurus, recording that its yellow glow was four times as bright as Venus. It remained visible for 653 days.

SHARP-EYED MAYANS

The Dresden Codex is a Mayan text produced in South America in the 11th or 12th century. It records with astonishing accuracy observations probably made 300 or 400 years earlier of the moon and Venus. Venus was the most important celestial body to the Mayans after the sun. The Mayans also seem to have been aware of the fuzzy nebula at the heart of the constellation of Orion: it featured in traditional stories and was represented on fireplaces with a smudge. They are the only civilization known to have discovered this feature of Orion without the use of telescopes.

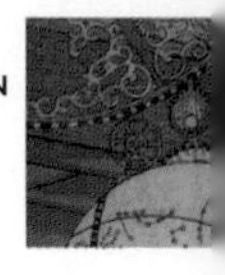

The Japanese poet Sadiae Fujiwara wrote about the star, and it was recorded in the pottery of Native American Anasazi and Mimbres artists. The "guest star" was the supernova that created the Crab nebula. After the disappearance of the new star from the night sky, it was not seen again for nearly 700 years, when the English doctor and astronomer John Bevis (1695–1771) discovered the nebula in 1731 using a telescope.

"God, when He created the world, moved each of the celestial orbs as He pleased, and in moving them he impressed in them impetuses which moved them without his having to move them any more... And those impetuses which he impressed in the celestial bodies were not decreased or corrupted afterward, because there was no inclination of the celestial bodies for other movements. Nor was there resistance which would be corruptive or repressive of that impetus."

Jean Buridan, 14th-century French philosopher

The Earth Moves—Again

Nearly two thousand years after Aristarchus first suggested that the Earth moves around the sun, the idea re-emerged. In the Christian world, it was a dangerous proposition, for the Church taught that the heavens were perfect and unchanging, that man was the pinnacle of creation and at the center of God's plan. How, then, could Earth have a subservient place, moving around the sun? The idea was heretical and immediately spelled trouble.

There were problems with the Ptolemaic model, the most significant of which was that the offset from Earth required for the focus of the moon's orbit was so great that the moon should be much closer to Earth at some times than at others—enough, in fact, to look noticeably larger. This problem, and other observations that cast Ptolemy's model into doubt, was revealed in 1496 by the German mathematician and astronomer Johannes Muller (1436–76), known by his Latinized name Regiomantus. The man who dared to challenge the Ptolemaic model was Copernicus—Mikolaj Kopernik—a Polish astronomer who did not trouble himself with observations but decided that it would be a neater solution if Earth orbited the sun rather than vice versa. Copernicus particularly disliked the small circles or mini orbits called "equants" that the planets needed to follow in the Ptolemaic model to explain their observed motions, and wanted a system in which there was a single, fixed center of the universe.

Copernicus

Although Copernicus completed his thinking on the sun-centered universe around 1510, he was cautious, and communicated it to only a few people before publishing his seminal work *De*

Revolutionibus Orbium Coelestium (On the Revolution of the Celestial Spheres) in 1543. The printer, Rheticus, was only part way through preparing Copernicus's book when he had to leave Nuremberg. The job passed to a Lutheran, Andreas Osiander, who added a preface in which he said that Copernicus did not mean the sun was *literally* at the center of the universe and he was presenting only a mathematical model that helped to explain observations. The preface was intended to defuse any criticism by the Church, but in fact the Catholic Church paid little attention to the book and only the Lutherans objected. Copernicus died the year the book was published and may never even have seen a copy. His book was largely ignored, and the print run of 400 copies did not even sell out, yet it has since been considered the text that started modern astronomy and helped spark the scientific revolution.

Though better than Ptolemy's spheres, there were still some problems with Copernicus's model. The fixed stars were thought to be on an invisible sphere beyond the farthest planet. For the stars not to appear to move, though, they needed to be very far away. We are comfortable with this concept today, but in the 16th century it immediately raised the question of why God would waste so much empty space between the farthest planet and the fixed stars. Another problem was that if the Earth was moving, why didn't the oceans slop about, and buildings get shaken to bits? On the other hand, unlike the Ptolemaic model, the Copernican model explained the observed movements of the planets without recourse to complex fudges.

Copernicus's explanation put the planets into two groups, with Mercury and Venus closer to the Sun than the Earth, then Mars, Jupiter, and Saturn farther out. (The other planets were unknown at the time.) Copernicus also worked out how long each planet took to orbit the sun, and the relative distances of the planets from the sun. These matched their grouping relative to the Earth's orbit, providing strong evidence in favor of his model.

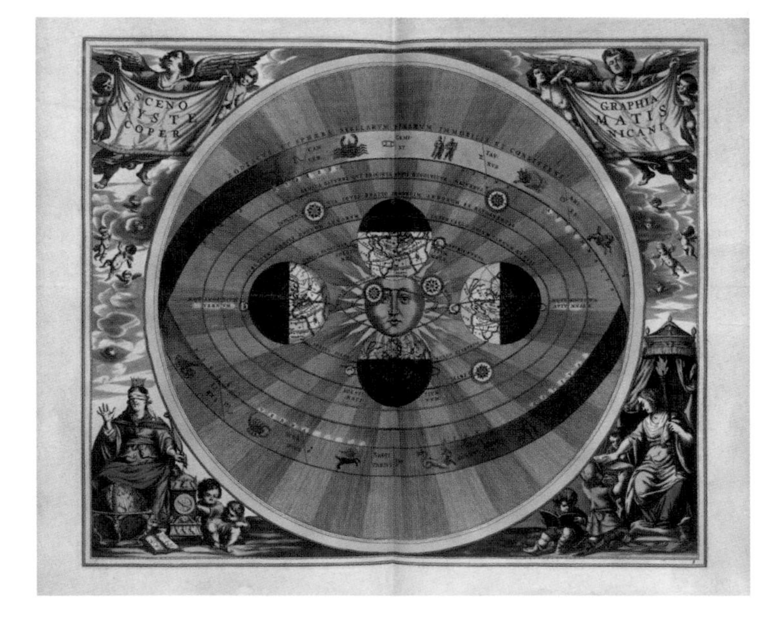

The Copernican model of the solar system, with planets orbiting the sun

EXPANDING UNIVERSE, SHRINKING EARTH

We all strive to put ourselves at the center of things. The disquiet that resulted from finding that Earth was not, after all, at the center of the solar system was immense. Nevertheless, astronomers assumed the solar system was important in the universe. Much later astronomers, who had recognized that the Milky Way is a galaxy, assumed the sun was near its center, and that the Milky Way was at the center of the universe—indeed *was* the universe. The discovery that the Milky Way is just a galaxy containing billions of stars, that the universe contains billions of galaxies, that the solar system is not central in the Milky Way galaxy, nor the Milky Way central in the universe dealt further blows to humanity's sense of self. We are, undoubtedly, insignificant beings on an insignificant spot of a planet in an ordinary solar system that's part of an ordinary galaxy—nothing special at all.

ALL CHANGE

Tycho Brahe was a colorful figure, an aristocrat kidnapped as a baby who later lost part of his nose in a duel and wore a gold and silver prosthesis thereafter. He became obsessed with the stars at an early age, and realized that consistent, accurate observations must form the foundation of any set of predictions. In 1569 he had a gigantic quadrant made, with a radius of around 20 feet (6 m). The rim was calibrated in minutes and allowed very precise measurements. He used it until it was destroyed in a storm in 1574.

In 1572, Tycho observed what appeared to be a very bright new star in the constellation of Cassiopeia. As the heavens were supposed to be fixed for all eternity, this was a cause of some consternation, and he set about recording its position over a period of months to determine whether it was a comet, which would move in relation to the fixed stars. He observed it for 18 months, during which time it faded from

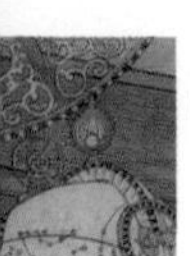

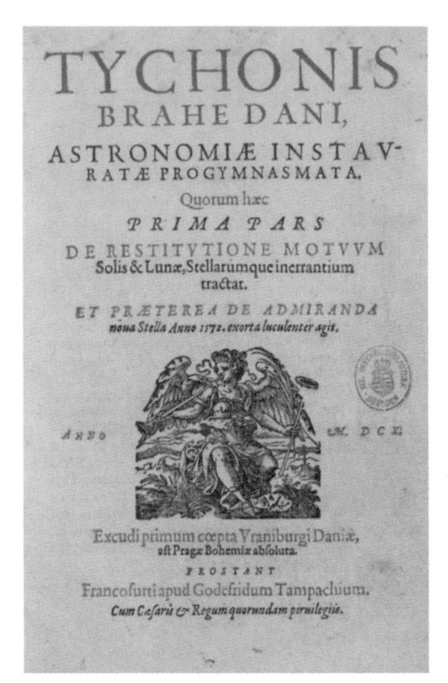

TYCHONIS
BRAHE DANI,
ASTRONOMIÆ INSTAV-
RATÆ PROGYMNASMATA.
Quorum hæc
PRIMA PARS
DE RESTITVTIONE MOTVVM
Solis & Lunæ, Stellarumque inerrantium
tractat.
ET PRÆTEREA DE ADMIRANDA
noua Stella Anno 1572. exorta luculenter agit.
ANNO M. DCII.
Excudi primum cœpta Vraniburgi Daniæ,
ast Pragæ Bohemiæ absoluta.
PROSTANT
Francofurti apud Godefridum Tampachium.
Cum Cæsaris & Regum quorundam priuilegiis.

Tycho Brahe's astronomical treatise showed his model of the solar system.

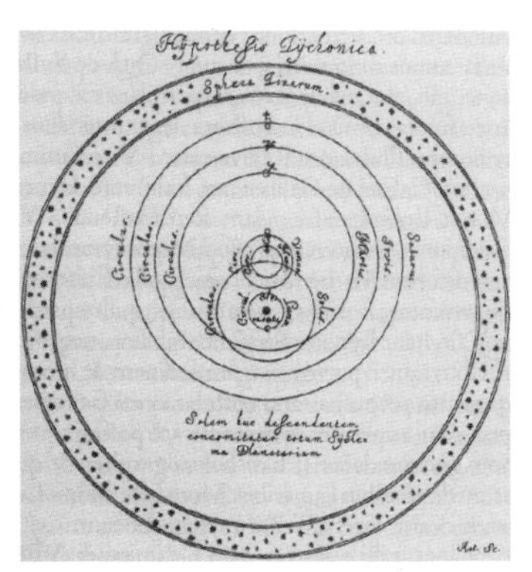

being brighter than Venus to appearing like a regular star, but did not change its position. When he published his account in *De Nova Stella* he gave astronomy a new term—nova. Tycho studied his data for evidence of the parallax that would be expected if Earth moved around the sun. Parallax is the apparent change of position of a nearby star against the background of more distant stars, when viewed from two different vantage points. As he found none, Tycho took his observations as disproving Copernicus's heliocentric model.

For all his scientific approach, Tycho was still of the opinion that events in the heavens presaged major changes on Earth and thought celestial phenomena were responsible for the religious wars going on at the time.

Nor would he countenance a moving Earth. If Earth were moving through space, he argued, a stone dropped from a tower would fall some distance from the foot of the tower because Earth would have moved on, leaving the stone behind. This was, of course, refuted by Gassendi in 1640 (see page 74).

A few years later, in 1577, Tycho made another Earth-changing observation, this time of a comet. His observations revealed that the comet could not be a local phenomenon, traveling very close to Earth and probably closer than the moon. Instead, it must travel between the planets. This meant that the Ptolemaic idea of crystal spheres hosting the planets and fixed stars had be to abandoned, as the comet would go crashing through them. It was almost as revolutionary, in its way, as the concept of a new star.

Tycho published his book in 1587–88, setting out his own model of the universe. It was a bit of a hybrid, keeping the Ptolemaic static Earth at the center of the universe, but having the other planets orbiting the sun, which was itself orbiting Earth.

It did away with the need for "deferents" and "epicycles" that had been required to make the Ptolemaic model work. Most importantly, though, it rejected the idea of crystal spheres and, for the first time, had the planets hanging unsupported in space.

Johannes Kepler (1571–1630)

A little younger than Tycho, Johannes Kepler was another prodigious astronomer but one who was forced to take a different approach. Kepler's enthusiasm for astronomy was sparked as a child when his mother took him to a high place to see the Great Comet of 1577 (the same one that prompted Tycho's work on comets). Kepler, though, was unable to make astronomical observations of his own as his eyesight was weak, having been damaged by smallpox. Instead, he applied mathematics to the study of the stars. Kepler trained as a priest, but his course at Tubingen, Germany, included mathematics and astronomy, in which he excelled. His tutor, Michael Maestlin, taught the Ptolemaic model officially, but in private introduced favored students—including Kepler—to Copernican astronomy.

Kepler was not independently wealthy, and one of the ways he earned extra money was by casting horoscopes. Unlike Tycho, who took seriously the link between earthly and celestial events, Kepler regarded horoscopes as total rubbish and privately referred to his clients as "fatheads." Still, it provided a useful income and kept him in favor.

Kepler developed his own model of the universe, published in 1597, which combined something of Copernicus with some rather arcane Greek physics in a bizarre fusion. Kepler suggested that the six planets (including Earth) occupied orbits that were defined by a set of spheres nested within and between the five geometric solids defined by Euclidean geometry. While this in itself was not to be particularly significant, he made a more important suggestion: that the planets were driven by a "vigor" emanating from the sun, but with reduced impact as distance from the sun increased. This was the first time that physical force had been cited as the source of the planets' motion, unless we count the idea that they were pushed along by angels.

Two for One on Astronomers in Prague

In 1597 Tycho moved to Prague to become the official Imperial Astronomer for the King of Bohemia and Holy Roman Emperor Rudolph II. It was here, in 1600, that Kepler met Tycho for the first time. While Tycho had amassed a prodigious amount of data, he did not have the mathematical skill to make the most of it. Kepler had the mathematical ability but no data on which to work. It looked like a perfect match, but their relationship was not an easy one. Having visited Tycho, Kepler returned to his family home in Graz, Austria, while Tycho was supposed to arrange funding for Kepler's work from Emperor Rudolph. Before negotiations were complete, Kepler and other Lutherans were evicted from Graz for refusing to convert to

Tycho Brahe

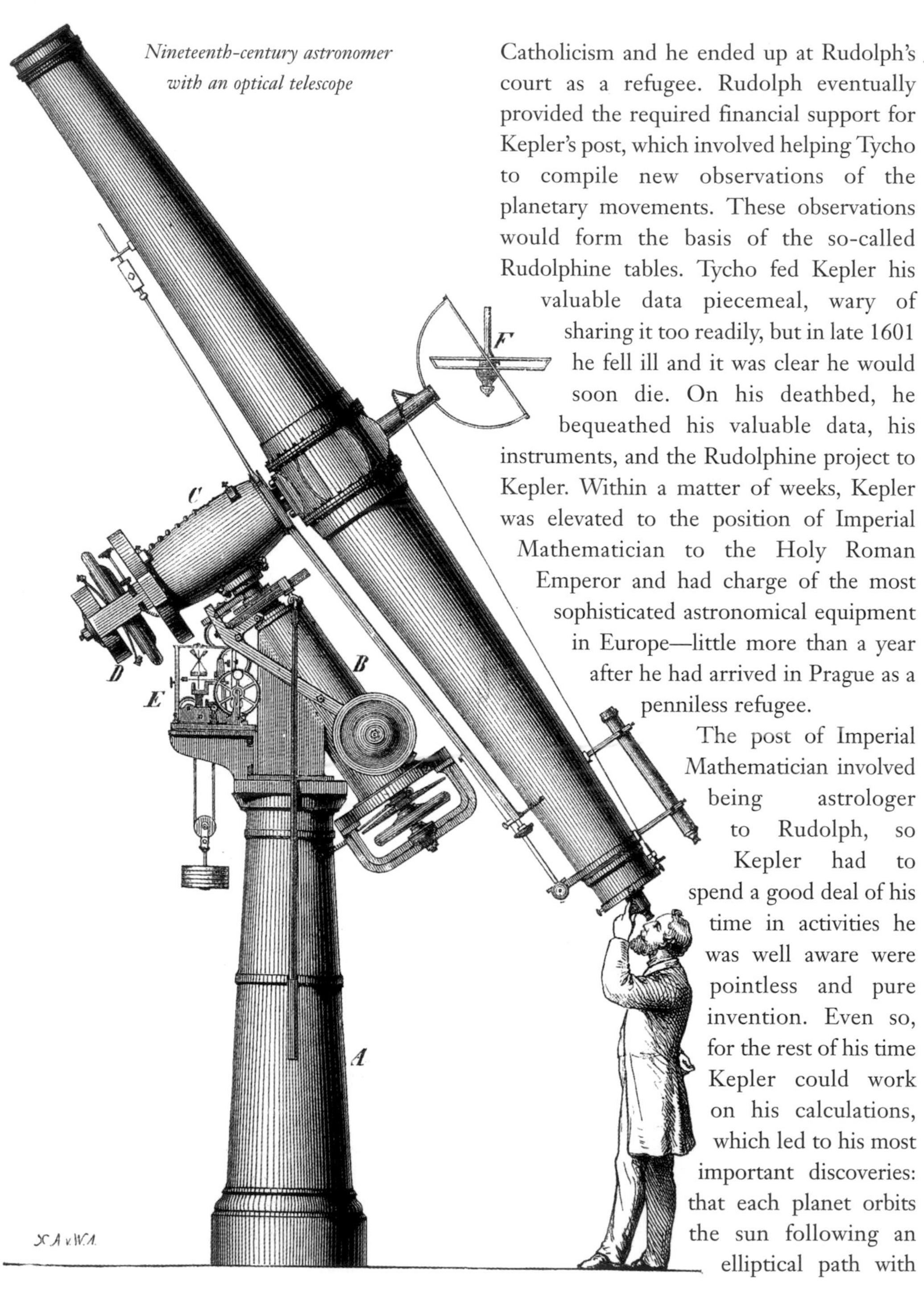

Nineteenth-century astronomer with an optical telescope

Catholicism and he ended up at Rudolph's court as a refugee. Rudolph eventually provided the required financial support for Kepler's post, which involved helping Tycho to compile new observations of the planetary movements. These observations would form the basis of the so-called Rudolphine tables. Tycho fed Kepler his valuable data piecemeal, wary of sharing it too readily, but in late 1601 he fell ill and it was clear he would soon die. On his deathbed, he bequeathed his valuable data, his instruments, and the Rudolphine project to Kepler. Within a matter of weeks, Kepler was elevated to the position of Imperial Mathematician to the Holy Roman Emperor and had charge of the most sophisticated astronomical equipment in Europe—little more than a year after he had arrived in Prague as a penniless refugee.

The post of Imperial Mathematician involved being astrologer to Rudolph, so Kepler had to spend a good deal of his time in activities he was well aware were pointless and pure invention. Even so, for the rest of his time Kepler could work on his calculations, which led to his most important discoveries: that each planet orbits the sun following an elliptical path with

the sun at one focus of the ellipse, and that the planets move more quickly when closest to the sun. Kepler's discoveries did not make him an overnight sensation, and in fact had relatively little impact. Many people still did not accept that Earth is not at the center of the universe. It was only when Isaac Newton took Kepler's work and explained, using gravity, why the planets have an elliptical orbit that the significance of his discovery really became apparent.

Religious turmoil, personal upheaval, and tragedy interceded to stall Kepler's work. His wife died (he later remarried) and then his mother was tried for witchcraft, although acquitted after spending several months in jail.

His third and final law, which came to him in 1618, describes how the square of the time it takes a planet to orbit the sun is proportional to the cube of its distance from the sun. For example, Mars is 1.52 times as far from the sun as the Earth is, and its year is 1.88 Earth years: $1.52^2 = 3.53 = 1.88^3$. The Rudolphine tables, finally published in 1627, were the first modern astronomical tables. They used the newly discovered logarithms developed by the Scottish mathematician and astronomer John Napier (1550–1617), which could be employed to determine the positions of the planets at any time in the past or future.

Achromatic telescope, mid-18th century (left); replica of Newton's reflecting telescope, 1672 (right)

The Invisible Becomes Visible

Whereas Tycho Brahe worked without a telescope, measuring the positions of stars and planets using compasses and quadrants, from 1610 Kepler had a telescope he could use—one sent to him by Galileo, to enable him to confirm Galileo's own observations. For astronomers, the world—and indeed the universe—changed with the invention of the telescope. Suddenly, the difference between stars and planets became apparent. Some planets were found to have their own moons, and the possibility that they could be other worlds emerged. The Milky Way resolved into a band of stars, and the stars truly became numberless.

The first astronomical telescope was made by Leonard Digges (1520–59) in England in the early 1550s but it did not come to public attention until his son

Thomas (1546–95) published his work on it in 1571, 12 years after Leonard's death. Thomas was only 13 when his father died and was passed into the care and tutelage of John Dee (1527–1609), the mathematician, philosopher, alchemist, and court astrologer to Queen Elizabeth I. This gave Thomas access to Dee's magnificent library, where he read Copernicus's book. In 1576, Thomas published his own most important work, a revised edition of his father's *Prognostication Everlasting*. He added not only an account of Copernicus's model of a universe centered on the sun, but his own theory that the universe is infinite. Rejecting the idea of the fixed stars on a distant sphere, Thomas Digges proposed an infinity of space in which stars continued forever. He cited no evidence for this theory, but it seems likely that his use of the telescope and the realization that the Milky Way is a band of stars brought him to this conclusion. Because Digges published in English rather than Latin, his ideas were accessible to many more people and the popularity of the Copernican model spread.

At about the same time, though, the Catholic Church began to take notice of the potentially heretical idea of a sun-centered universe. The source of their animosity seems to have been that the model was supported by Giordano Bruno, burned at the stake for heresy in 1600. Bruno was a follower of a religious movement called Hermetism, based on ancient Egyptian beliefs that the sun is a god and should be worshiped. He was naturally attracted to a heliocentric model of the universe. His promotion of Copernicus's model attracted the attention of the Church, but the popular belief that he was burned for his promotion of the Copernican model is without foundation. He was actually condemned for believing that Christ was created by God rather than being God (Arianism) and for the practice of magic. However, Bruno's support of the heliocentric model increased the Church's hostility toward it and by extension to Digges' theory of the infinite universe. Despite his rather wacky religious ideas, Bruno had insights far ahead of his time as an astronomer. He suggested that the distant stars might be just like our sun, might have worlds of their own, and that these might even be home to beings as glorious as humankind.

GALILEO IN SPACE

NASA launched a spacecraft named after Galileo in 1989, which went into orbit around Jupiter in 1995. En route, the Galileo craft passed through the asteroid belt where it discovered a miniature moon, called Dactyl, in orbit around the asteroid Ida. In 1994, Galileo photographed fragments of the comet Shoemaker-Levy 9 as it crashed into Jupiter. A probe released into Jupiter's atmosphere recorded winds of around 450 mph (720 kph) before being destroyed by the Jovian atmosphere. Galileo made 11 orbits, recording data about the planet and its moons on its primary mission. The craft's mission was extended and it studied Jupiter's volcanic moon Io and its icy moon Ganymede. Galileo was deliberately destroyed in 2003, being burned up in the atmosphere of Jupiter.

The asteroid Ida with its tiny moon, Dactyl. Ida is 35 miles (56 km) long, and Dactyl only 1 mile (1.6 km) long.

Galileo, Master of the Universe

The greatest early user of the telescope was undoubtedly Galileo. Galileo turned his attention to astronomy in 1604, studying the supernova that Kepler had observed. He established that it did not move and so must be as far away as the other stars. Galileo made his own telescopes, which were very powerful for the time (see page 41). In 1610 he had an instrument with a magnifying power of 30, with which he first observed the four brightest moons of Jupiter (now called the "Galilean moons"). (The largest of Jupiter's moons, now called Ganymede, was apparently spotted by Chinese astronomer Gan De in 364 BCE with the naked eye.) At first Galileo thought they were "fixed stars" near to Jupiter, but repeated observation showed that they move. When one disappeared, he realized it had gone behind Jupiter and so must be orbiting the planet. These were the first bodies identified as orbiting something other than the sun or Earth and the impact on contemporary cosmology was immense. No further Jovian moons were found until 1892, though there are now 63 known moons with a relatively stable orbit around the planet, and more small moons may yet be found.

Also in 1610, Galileo observed the phases of Venus (similar to phases of the moon). This proved conclusively that the planet must orbit the sun and that the phases are due to the way various parts are illuminated by the sun during the stages of its orbit. As a result, most astronomers switched their allegiance from the Ptolemaic to one of the heliocentric models of the universe during the early 17th century.

Yet this was not all. Galileo also observed Saturn's rings, although he was not able to work out what they were. He realized that the Milky Way is actually a band of a massive number of stars, saw that the moon has craters and mountains, observed sunspots, and distinguished between planets and stars. He stated that stars are distant suns, and made estimates of their distance from Earth based on their relative brightness. Although he put the closest stars at only several hundred times the distance of Earth from the sun, and those visible with a telescope at several thousand times the Earth–sun distance (well short of the real distances, of course), these figures made a mockery of the arguments against the Copernican model that the stars could not be very distant. He made clear, too, that the stars are not all at a fixed distance, but spread throughout space. In *Sidereus Nuncius* (Starry Messenger) published in 1610, he stated that the planets resolve into disks when viewed through the telescope, while the stars remain points of light. He observed Neptune, but did not realize it was a planet. He even identified sunspots, which had also been seen by the German astronomer Johann Fabricius (1587–1616) and the English astronomer Thomas Harriot (1560–1621), and concluded that the sun revolves on its own axis once every 25 days. The sunspots were going to be more significant for Galileo's life than they warranted.

> ***EPPUR SI MUOVE***
>
> It is commonly said that Galileo, after renouncing his belief that Earth moves around the sun, mumbled *"eppur si muove"*—"yet it moves." The earliest source for this is a century after his death, and it's unlikely he would have done something so provocative within hearing of the Inquisition.

Crossing Swords with God

Galileo's observations provided ample evidence in favor of the Copernican model of a heliocentric solar system and a moving Earth, but Galileo shied away from public support for this model, made anxious no doubt by the fate of Giordano Bruno. At first, the Church was interested and even enthusiastic about Galileo's discoveries. He visited Pope Paul V in 1611, and a subcommittee of Jesuit priests endorsed his findings that the Milky

Galileo observed sunspots (dangerously!) with his telescope in 1612.

Pope Paul V (1552–1621)

BESTSELLER OF 1610

Galileo sent an advance copy of his book *The Starry Messenger* to the court at Florence on March 13, 1610. By March 19, the whole print run of 550 copies had sold out. The book was translated into many other languages immediately, and within five years it was even available in Chinese!

Way is a vast collection of stars, Saturn has a strange oval shape with lumps on each side (they were not identified as rings), the moon has an irregular surface, Jupiter has four moons, and Venus has phases. The committee did not comment on the implications of the findings. While in Rome visiting the pope, Galileo became a member of one of the first scientific societies in the world, the Lyncean Academy, and at a banquet in his honor the name "telescope" was first suggested for the new astronomical instrument.

Galileo's good relationship with the Church was not to last, though. He produced a pamphlet on sunspots in which he made his only published statement in favor of the Copernican model. It attracted the attention of the Church, and when he visited Rome in 1615 the papacy set up an inquiry into Copernican beliefs and concluded they were "foolish and absurd… and formally heretical." Soon after, Galileo was told he must not hold, defend, or teach Copernican beliefs and that he would face the Inquisition if he disobeyed. He heeded the warning, at first. In 1629, Galileo wrote his *Dialogue of the Two Chief World Systems* which presented the Copernican and Ptolemaic models in the form of an imaginary dialogue between defenders of each system. He published with the permission of the Church, on condition that he would not favor Copernicanism. The papal censor insisted on a preface and a final statement saying the Copernican view was given as a hypothesis and saying Galileo could change the wording of these as long as their substance remained the same. The changes Galileo made to the preface, and the fact that the character in the book called Simplicio who supports the Ptolemaic model is clearly a simpleton, led Pope Urban VIII to believe Galileo was making

BEHIND THE TIMES

The *Dialogue* and Copernicus's *De Revolutionibus* remained on the Catholic Church's Index of Banned Books even after the general prohibition on books teaching heliocentrism was lifted in 1758. As late as 1820, the Church's censor refused a licence to a book that treated heliocentrism as an established fact. An appeal against the decision had it overturned, and both Galileo's book and Copernicus's book were removed from the Index at the next publication in 1835. The Catholic Church eventually apologized for its treatment of Galileo—but not until 2000. Pope John Paul II cited Galileo's trial among other errors committed in the previous 2,000 years that the Church was owning up to rather belatedly.

HALLEY AS CATALYST

When Halley visited Newton in Cambridge in 1684, the two talked about an idea that senior astronomers had already been discussing for a while—the relation of the inverse square law to the attraction that keeps planets in orbit. Halley had discussed it with Robert Hooke and Christopher Wren in January of the same year. Halley asked Newton what he thought the orbit of a planet would be if the force between it and the sun were reciprocal to the square of its distance from the sun. Newton answered that he had already calculated it and that it would be an ellipse. As a result of this conversation, Newton went on to publish the *Principia*, finally releasing work he had sat on for years. It became the most important scientific text ever published.

fun at his expense and promoting Copernicanism. Galileo was summoned to Rome to stand trial for heresy—for "holding as true the false doctrine taught by some that the sun is the center of the world." Galileo was persuaded to plead guilty to avoid the Inquisition and possible torture. He agreed that he had gone too far in presenting the case for Copernicanism.

His punishment was life imprisonment, which eventually took the form of house arrest in his own home from 1634 until his death in 1642.

During the last years of his life, Galileo wrote his greatest work, *Discourses and Mathematical Demonstrations Concerning Two New Sciences*.

The first modern scientific textbook, it spelled out the scientific method and gave mathematical or physical explanations for phenomena that had previously been dealt with using only the tools of philosophy. The book was smuggled out of Italy and published in Leiden in Germany in 1638. It was hugely popular and influential everywhere except Italy.

Cataloging the Skies

The development of the telescope allowed astronomers to make far more accurate maps of the stars. Prompted by rivalry with the French, who had set up a national observatory under the control of the French Academy, the Royal Society of London pressed for the foundation of an observatory in Britain. The Royal Observatory was established at Greenwich in 1675, with John Flamsteed (1646–1719) as the first Astronomer Royal (though the title then was "Astronomical Observator"). Flamsteed

was soon in correspondence with the young Edmund Halley (1656–1742), then a student at Oxford and already a keen astronomer—he took a telescope of more than 23 feet (7 m) in length with him to Oxford University. Halley first wrote to Flamsteed with suggested corrections to the star catalog then in use, and soon became something of a protégé to Flamsteed. Flamsteed was engaged on making a new catalog of stars of the northern hemisphere. Halley proposed a parallel study in the southern hemisphere and soon secured royal approval. Halley's father financed it, giving his son an allowance that was three times Flamsteed's royal salary.

Seeing More and More

As the power of telescopes continued to improve, astronomers could reveal more and more about the mysteries that puzzled earlier scientists. Galileo had discovered Saturn's "ears," which then strangely disappeared a few years later. In 1655, Huygens began working with his brother Constantijn on an improved telescope design that prevented chromatic aberration—the colored fringes around images. He then turned his 50 times magnification telescope on Saturn. In 1652 he discovered Saturn's largest moon, Titan, and four years later saw that the "ears" that Galileo had seen on Saturn were actually a ring: "…the planet is surrounded by a thin

SLOW RETURN

Newton's comet, the Great Comet of 1680, was the first comet to be observed with a telescope. It is set to return around 11037. Newton used his measurements of the comet's path to test Kepler's laws.

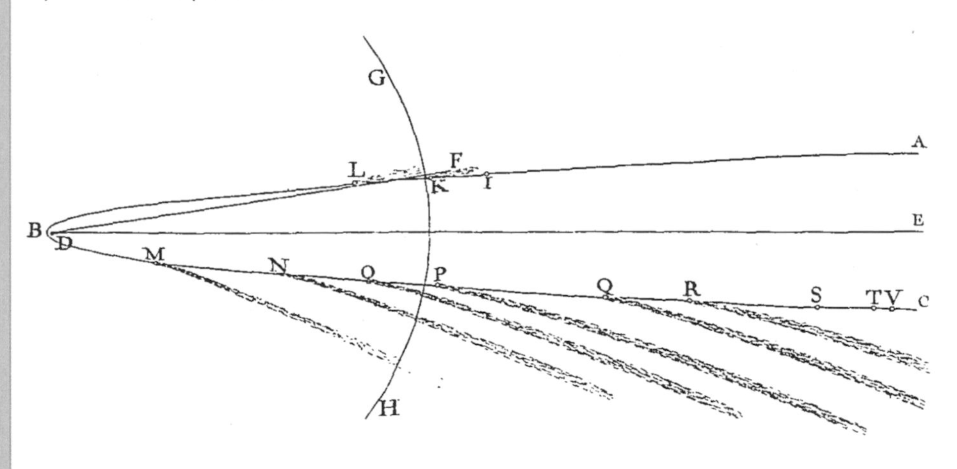

Newton's diagram of the orbit of the comet of 1680, showing its parabolic path

The alignment of the sun, Earth, and Mars gave 17th-century astronomers an opportunity to calculate the size of the sun and its distance from the Earth.

flat ring, nowhere touching, and inclined to the ecliptic." It wasn't clear what the ring was made of, though. At first, astronomers assumed it was solid or liquid, but in 1675 Giovanni Cassini discovered a gap in the system of rings. Deciding the nature of the ring was set as the topic of the Adam's Prize Essay at Cambridge University in 1855. It was won by James Clerk Maxwell, who demonstrated that a collection of tiny orbiting solid particles is the only possibility as anything else would be unstable; only the distance from Earth to Saturn made the system look like a continuous mass. Maxwell was proved correct in 1895, using spectroscopic techniques.

Far, Far Away

Cassini is most famous for his work on the distance between planets and the size of the sun. Before this, the only estimates of the distance of the sun from Earth was that furnished by Aristarchus in 280 BCE. Copernicus's work made it possible to judge the ratios of distances from each planet to the sun, but there was no figure from which to calculate the absolute distances. A perfect opportunity presented itself in 1671 as the sun, Earth, and Mars were aligned and the distance between Earth and Mars was at its minimum. As director of the Paris Observatory, which opened that year, Cassini was able to send a colleague, Jean Richer, to Cayenne in South America to make observations, while he made his own observations in Paris. As the king at the time was Louis XIV—the Sun King—the project met with royal approval. Knowing that around 6,200 miles (10,000 km) separated Paris and Cayenne, Cassini used trigonometry to calculate the distance of Mars from the Earth, then applied Kepler's laws of planetary motion to deduce that the sun was 86 million miles (138 million km) from Earth. This is only 9 percent less than the currently accepted figure of nearly 150 million km (93 million miles). Further calculations revealed that the sun is 110 times the size of Earth. After the publication of Newton's *Principia* and its description of gravity, it became clear that the sun is about 330,000 times the mass of Earth.

Putting Comets in their Place

The friendship between Halley and Newton bore fruit in the form of an explanation of the motion of comets. Newton showed in the *Principia* how the path of a comet could

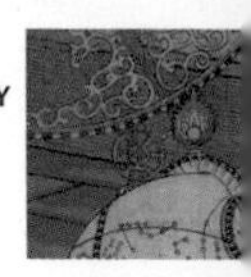

TRANSIT OF VENUS

Before Cassini, the English astronomer Jeremiah Horrocks (1618–41) suggested that by accurately timing the transit of Venus—the planet's passage across the face of the sun—from different places on Earth it would be possible to calculate the Earth–sun distance. Horrocks himself had observed a transit of Venus in 1639, two years before his death. The next was due in 1761, and again in 1769. Halley popularized the idea of using triangulation to calculate the Earth–sun distance, known as one astronomical unit (AU), which could be used to calculate the size of the solar system as then known. Triangulation is a way of calculating the position of something by measuring the angle to it from two fixed points a known distance apart. The method was traditionally used to measure the height of buildings and even mountains.

Halley died 19 years before the next transit would take place and so it was left to others to put his idea into effect. As the date approached, astronomers set off on expeditions around the world to record the timings. The transit proved very difficult to measure accurately and reliably, but by putting several different measurements together taken from different parts of the globe they arrived at a figure of around 95 million miles (153 million km), not far from today's currently accepted figure of 93 million miles (150 million km). By the end of the 18th century, then, astronomers had a realistic idea of the size of the solar system. The foundations were laid for the modern age of astronomy, an age in which the most distant celestial bodies would come into focus.

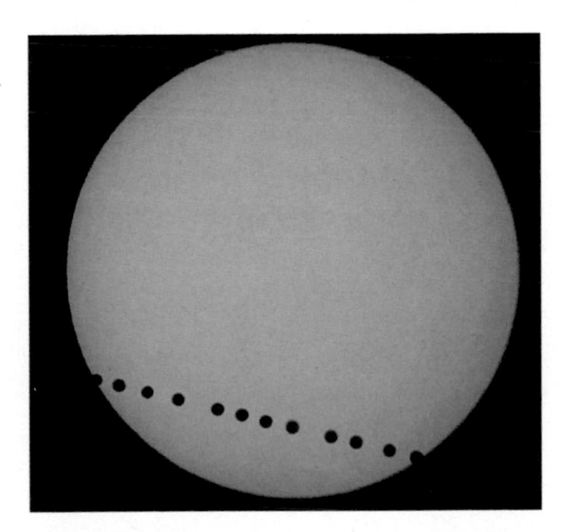

During a transit of Venus, the planet appears as a tiny dark spot passing in front of the sun.

be calculated from three observed positions over a period of two months and he compiled data on 23 comets. He assumed, though, that comets followed a parabolic path, coming from outside the solar system, looping around the sun, and heading off into outer space again—one which would now be considered a non-periodic comet. Unwilling to carry out the calculations on his comet data, Newton handed the figures to Halley. He, too, assumed the path was parabolic until he noticed that the path of the 1607

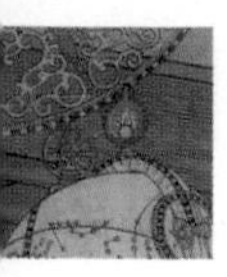

comet (observed by Kepler) was very similar to that of the 1680 comet he had seen himself. Later he found that it also matched the path of a comet seen in 1531 and concluded that all three were the same object, not following a parabolic path but a very wide elliptical orbit around the sun. Halley predicted the reappearance of the same comet in 1758, having calculated a 76-year return period. The comet—now known as Halley's comet—duly reappeared on Christmas Day 1758, 16 years after Halley's death.

Halley's Comet in History

Halley's comet may have been recorded as long ago as 467/6 BCE in both Ancient Greece and China. A meteor the size of a "wagonload" that fell while the comet was in the sky remained a curiosity and an attraction in Greece for 500 years. The first certain record of Halley's comet is Chinese, from the apparition in 240 BCE. The next sighting, in 164 BCE, is recorded on a Babylonian clay tablet. Coins depicting the Armenian king Tigranes the Great seem to show Halley's comet on his crown, recording its appearance in 87 CE. It made

The first passage of Halley's comet was photographed in 1910.

The Bayeux tapestry shows Halley's comet appearing in 1066, when it was taken as an omen.

THE COMET BRINGETH AND THE COMET TAKETH AWAY

"I came in with Halley's Comet in 1835. It is coming again next year, and I expect to go out with it. It will be the greatest disappointment of my life if I don't go out with Halley's Comet. The Almighty has said, no doubt: 'Now here are these two unaccountable freaks; they came in together, they must go out together.'"

Mark Twain, autobiography, 1909

Twain was born on November 30, 1835, exactly two weeks after Halley's comet made its closest approach to the sun (perihelion). He died on April 21, 1910, the day following the comet's next perihelion.

its closest approach in 837 CE, at a distance of only 0.03 AU, when its tail may have extended as much as 60 degrees across the sky. Halley's comet is depicted in the Bayeux tapestry, and possibly in Giotto's *Adoration of the Magi* as the star of Bethlehem (which it probably wasn't, as it appeared in 12 BCE).

The comet appeared spectacularly in 1910 with a relatively close approach of 0.15 AU. It was photographed for the first time, and its tail studied by spectroscopy (a method of analyzing the chemical composition of a gaseous body by studying the characteristic pattern of spectral lines it produces, see page 116). Its spectrum revealed (among other things) that the tail contained the toxic gas cyanogen. This prompted the astronomer Camille Flammarion (1842–1925) to say that passing through the tail would "possibly snuff out all life on [Earth]." As a result, the public was duped into spending a fortune on gas masks, "anti-comet pills," and "anti-comet umbrellas." Needless to say, life on Earth survived the encounter.

The comet's return in 1986 prompted not only photography from Earth but close-up inspection in space by two probes, Giotto and Vega. These found that the comet is shaped rather like a peanut, is 9 miles (15 km) long and 5 miles (8 km) wide and thick, with a coma (or atmosphere) 62,000 miles (100,000 km) across. The coma forms as solid carbon monoxide and carbon dioxide on its surface turn into gas (sublime) with the sun's rays. Comet Halley is thought to be composed of small pieces, called a rubble pile, held loosely together. They rotate as a body about every 52 hours. The two probes mapped about a quarter of the comet's surface, finding hills, mountains, ridges, depressions, and a crater.

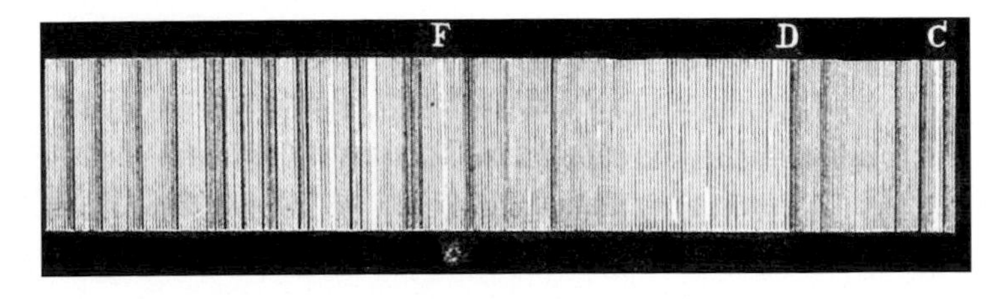

Variable star spectrum for the constellation Corona borealis (northern crown), 1877

Williamina Fleming

Spectroscopy—A New Way of Viewing

At the end of the 19th century there emerged an entirely new way of looking at stars, by studying their spectrum using a technique called spectroscopy. As light passes through a gas some wavelengths are absorbed, leaving a characteristic pattern of spectral lines. Each gas creates its own unique spectral pattern. So by analyzing the light from a star it is possible to work out its chemical composition. The American astronomer Henry Draper (1837–82), pioneer of astrophotography, was the first to photograph a star's spectrum, in 1872. His photographs of Vega showed distinct spectral lines. He took more than 100 photographs of star spectra before his death in 1882. In 1885, Edward Pickering (1846–1919) picked up the baton, and began to supervise the wide-scale use of photographic spectroscopy as director of Harvard College Observatory to produce a detailed star catalog. Draper's widow agreed to fund the venture, and the ambitious cataloging that would eventually yield the Henry Draper Catalog was begun. The first publication was the *Draper Catalog of Stellar Spectra* in 1890, which classified 10,351 stars.

Pickering became frustrated with the competence of his male assistants and declared that his maid could do a better job. His maid was a Scottish woman, Williamina Fleming (1857–1911), who had emigrated with her husband but had then been abandoned while pregnant. She had gone to work for Pickering to support herself and her son. Fleming took on the task of cataloging and classifying stars, developing a system of assigning them a letter according to how much hydrogen is in their spectra (with A for the most). In nine years,

Annie Jump Cannon

Fleming cataloged over 10,000 stars. She discovered 59 gaseous nebulae, over 310 variable stars, 10 novae, and the Horsehead nebula. Pickering put her in charge of a large team of women called "computers," whom he employed to carry out the necessary calculations involved in classifying and cataloging the stars. (The women were paid only 25–50 cents an hour, less than secretaries received at the time.) Fleming and several other women in the team, including Henrietta Swan Leavitt (1868–1921) and Henry Draper's niece, Antonia Maury (1866–1952), became respected astronomers in their own right.

Another of "Pickering's women" was Annie Jump Cannon (1863–1941), who improved on Fleming's system and introduced classification of stars based on temperature. Unlike Fleming, Cannon had a degree in physics and was already studying astronomy when she started working for Pickering. She was almost completely deaf following a bout of scarlet fever, yet it was she who negotiated when Maury and Fleming argued about classification methods. Cannon's new method classified stars as O, B, A, F, G, K, M (the usual mnemonic is "Oh, Be A Fine Guy/Girl, Kiss Me"), a system known as the Harvard spectral classification scheme that is still in use today. A refinement of the scheme, called the Morgan-Keenan system, supplements each letter with numbers 0–9 to fine-tune it, and adds Roman numerals I to V to indicate luminosity, but Cannon's system remains at its heart. Cannon would later take over the cataloging project.

With all its supplements, the Draper catalog has recorded and classified 359,083

PARALLAX

Parallax is a method of calculating the distance to an object by observing the object from two different positions. In the case of a star, the sky is photographed twice, six months apart. By measuring how far the star seems to have moved in relation to the background stars, astronomers can use triangulation to work out the distance from earth to the star.

You can see how the principle of parallax works by holding up a pencil in front of your face and looking at it first with the left eye only and then the right eye only. The pencil seems to move in relation to the background because each eye views it from a slightly different position.

stars. Cannon personally classified 230,000 stars, more than all previous astronomers put together. She was the first woman to be awarded an honorary doctorate by Oxford University and the first woman to be elected an officer of the American Astronomical Society.

Looking into the Void

The triangulation method that Cassini used in the 17th century to estimate the distance to Mars could, with skill, be used to estimate the distance to nearby stars. It means using the Earth's positions six months apart—that is, on either side of the sun—to provide the baseline for the triangulation. As the Earth–sun distance is one AU, this baseline will be two AU wide, a large enough distance for the accurate measurements required. During this time a nearby star will be seen to have changed position compared with the more distant background stars—a method known as parallax (see panel, page 171).

Huygens had earlier tried to estimate the distance of Sirius from Earth by comparing its brightness with that of the sun. He decided that, assuming Sirius were as bright as the sun, it would be 27,664 times farther away. It was a difficult task, as he had to compare his observations of the sun made in daytime with his observations of Sirius seen at night.

Although the principle of measuring the apparent movement of a star across the sky to calculate its distance is sound, the technique was difficult and required equipment that was simply not available to early astronomers. The first accurate stellar distance discovered by parallax was calculated by the German scientist Friedrich Bessel (1784–1846) who, in 1838, worked out a distance of 10.3 light years for 61 Cygnus. In fact, a Scot, Thomas Henderson (1798–1844), had already measured the distance to Alpha Centauri in 1832 but did not publish his results until 1839. With the distance to a star known, it's relatively straightforward to reverse Huygen's equations to calculate its brightness.

Yet still the tools available were not really up to the task. Measurements had to be made by eye and photography had yet to be invented. By 1900, only 60 parallaxes had been measured. With the advent of photography, the process could be speeded up dramatically, and the next 50 years yielded a further 10,000 parallaxes. Between 1989 and 1993, the European Space

The Hipparcos satellite was used for measuring the parallaxes of over 100,000 stars.

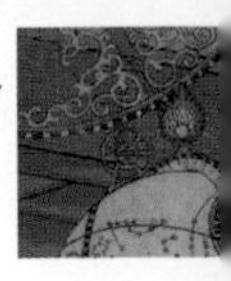

TELESCOPES IN SPACE

The Hubble Space Telescope, launched using the Space Shuttle in 1990 and named in honor of the famous astronomer, is an optical telescope in orbit around the Earth. Because it is in space, it produces images of extreme clarity with almost no interference from background light nor distortion from Earth's atmosphere. Space telescopes were first proposed in 1923, long before it became possible to build one.

In this Hubble image, two galaxies are being drawn together by their mutual gravitational attraction.

Agency's Hipparcos satellite measured the parallaxes of 118,000 stars, and the Tycho-2 catalog from the same mission provides data for more than two and a half million stars in the Milky Way.

For very distant stars parallax is of little use. Another method, using data from stars called Cepheids, was developed by Henrietta Swan Leavitt, one of Henry Pickering's team of female "computers." Cepheids vary in intensity, pulsating at intervals of anything from one day to hundreds of days. Once the distance to one Cepheid had been calculated, Leavitt's equation relating period-luminosity to distance meant that the distance of other Cepheids could be worked out. Suddenly, distances across and even outside the Milky Way became apparent, and the universe was found to be much bigger than thought.

In 1918 the American astronomer Harlow Shapley (1885–1972) used the Cepheid method to study globular clusters, which he thought were inside the Milky Way. He realized that the Milky Way was very much bigger than previously thought and that the solar system was not even near the center, as had been assumed. In the

winter of 1923–4, the American astronomer Edwin Hubble (1889–1953) found Cepheids outside the Milky Way, in the Andromeda Galaxy, and was able to calculate the distance to the galaxy at about one million light years (his figure was low, it is actually about two and a half million light years away).

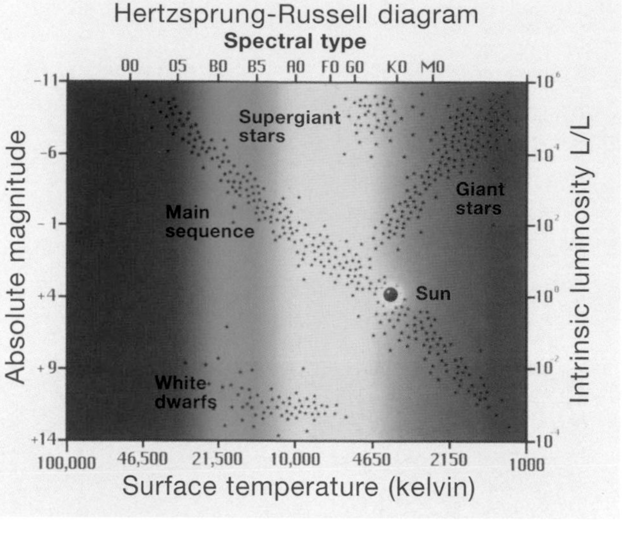

Hertzsprung-Russell developed this diagram showing the brightness (y-axis) and temperature (x- axis) of stars; color changes with temperature.

Stripes from Stars

The Danish chemical engineer Ejnar Hertzsprung (1873–1967) was studying astronomy and photography in his spare time when he discovered a relationship between the color of a star and its brightness. Although Hertzsprung eventually became a renowned professional astronomer, he was still an amateur when he published his results in 1905 and in 1907 in an unremarkable photographic journal. His discovery went unnoticed by professional astronomers. The American astronomer Henry Norris Russell (1877–1957) also noted the relationship between stellar brightness and color, but he published his discovery in a better-known astronomical journal in 1913. In addition, Russell plotted

Ejnar Hertzsprung

Henry Russell

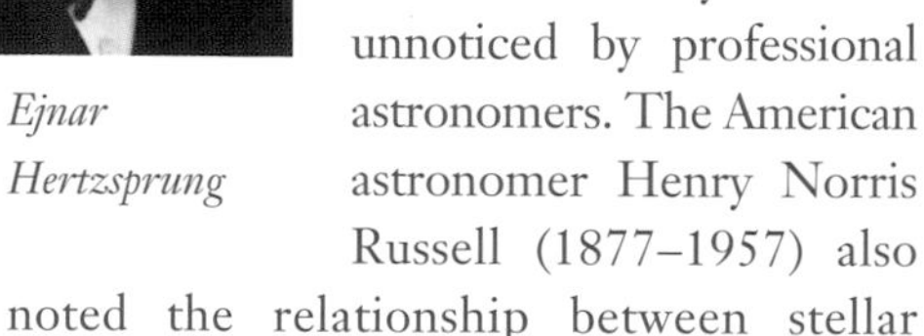

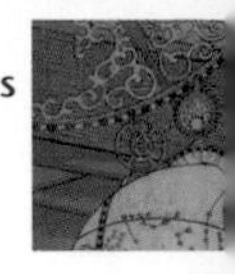

his results as a graph. Hertzsprung's contribution was recognized early on and the graph is now known as the Hertzsprung-Russell diagram.

The color of a star—or more precisely, the wavelength of the light it emits—is an indication of its temperature. Yet the overall brightness of a star depends also on its size. Just as a room heater may emit more heat than a (much hotter) burning match, the size of a star is as important as its temperature. So a huge red star may emit more energy than a small blue star, even though the surface temperature of the blue star is higher. Information from the Hertzsprung-Russell diagram gave astronomers the first inkling of what might be happening inside stars.

The Secret Life of Stars

Arthur Eddington, the British astronomer who led the expedition to observe the solar eclipse in 1917 that confirmed Einstein's theory of relativity, gleaned the first insight into what might be happening within a star. By combining information from the Hertzsprung-Russell diagram and the known mass of some stars, he found that the most massive stars are the brightest. This makes sense. In order to prevent gravity pulling the star in on itself, it must produce and emit a lot of energy. The larger the mass, the greater the pull of gravity and the more energy needed to resist it. He soon discovered that, regardless of size and surface temperature, the internal temperature of all main sequence stars is about the same. He realized, too, that the fuel providing the energy for a star must be nuclear—there was no other way a star could have a large enough supply of fuel to keep burning for billions of years.

Arthur Eddington

The first suggestion was that the sun's energy was derived from radioactive isotopes such as radium, but the half-life of radium is too short. The important breakthrough came through work carried out at the Cavendish atomic research center in Cambridge, England. In 1920 the British chemist and physicist Francis Aston (1877–1945) used a mass spectrometer to measure the mass of hydrogen and helium. The hydrogen nucleus has one proton,

Mass spectrometer used for measuring stable carbon and oxygen isotopes

> *"A star is drawing on some vast reservoir of energy by means unknown to us. This reservoir can scarcely be other than the sub-atomic energy which, it is known, exists abundantly in all matter; we sometimes dream that man will one day learn to release it and use it for his service. The store is well-nigh inexhaustible, if only it could be tapped. There is sufficient in the Sun to maintain its output of heat for 15 billion years."* Arthur Eddington, 1920

whereas the helium nucleus has two protons and two neutrons. Aston discovered that four hydrogen nuclei had slightly more mass than one helium nucleus. Eddington knew that hydrogen and helium were by far the most abundant elements in the sun. With his knowledge of Einstein's work, Eddington was able to apply the equation $E=mc^2$ to the sun and deduce that its energy came from nuclear fusion, hydrogen being forged into helium at the heart of the sun. The slight difference in mass that Aston had noted would be turned into energy.

Just as nuclear fission transforms heavier elements into lighter elements by breaking the nucleus apart, so nuclear fusion transforms lighter elements into heavier elements by combining nuclei. The huge volume of gas involved meant there was enough energy being released to power the sun for billions of years. Later it was realized that all the elements other than hydrogen, helium, and some lithium were formed by fusion inside stars or supernovae.

Listening to the Void

Although we already deal in distances and numbers of stars unimaginable to the early stargazers, there is yet more that we cannot see with optical telescopes, even those anchored in space. But by making use of the non-visible parts of the electromagnetic spectrum, such as radio waves, it has been possible to probe ever deeper into the cosmos.

Perhaps the origins of radio astronomy lie with the inventor and entrepreneur Thomas Edison (1847–1931), who suggested in a letter written in 1890 that he and a colleague might build a receiver to pick up radio waves from the sun. If he ever built such a device, it would not, sadly, have detected radio waves from space. The British physicist Sir Oliver Lodge (1851–1940) actually built a detector, but failed to find any evidence of radio waves from the sun in 1897–1900. The first

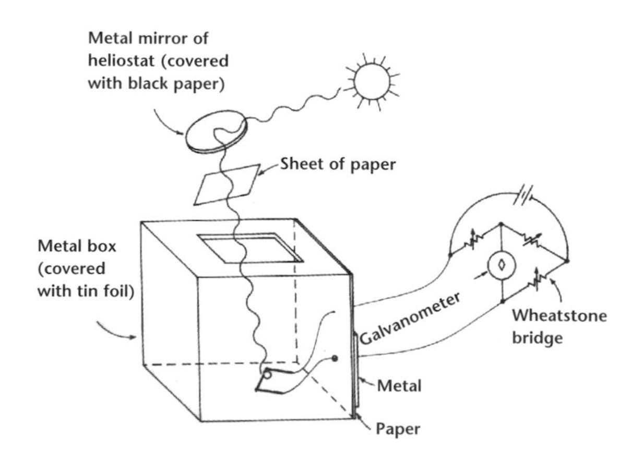

Wilsing and Scheiner's equipment attempted to detect radio waves from the sun.

NIKOLA TESLA (1856–1943)

Nikola Tesla was born in the Austro-Hungarian Empire, in an area which is now part of Croatia. He dropped out of university twice and cut all links with his family and friends (his friends believed he had drowned in the River Mura). In 1884, he moved to the USA.

Tesla worked on wireless communication, X-rays, electricity, and energy. When he arrived in the USA he began working for Thomas Edison but resigned over a dispute about pay. He later set up his own laboratory. He was a very prolific inventor, but some of his inventions, his character, and his attitude were eccentric, and he has always been considered a maverick. His claims to have detected radio transmissions from aliens on Mars or Venus did not help.

In 1904 the US Patent Office took away Tesla's patent for radio and gave it to Marconi instead; Marconi was awarded the Nobel Prize for the invention of radio in 1909. After squabbles with Marconi and Eddington, and the demolition of his Telefunken wireless station at Long Island by the navy in case it were used for spying in the First World War, Tesla's fortunes took a turn for the worse. He became increasingly obsessed with the number three and with pigeons.

The last nail in the coffin of his reputation was his promotion of his so-called "death ray" which would, he claimed, "send concentrated beams of particles through the free air, of such tremendous energy that they will bring down a fleet of 10,000 enemy airplanes at a distance of 200 miles ... and will cause armies to drop dead in their tracks." Tesla lived the last 10 years of his life in the Hotel New Yorker, and when he died two truckloads of his papers were impounded by the US government as a security risk.

scientists to look into the issue in depth were the astronomers Johannes Wilsing (1856–1943) and Julius Scheiner (1858–1913), working in Germany. They concluded that radio astronomy fails because radio waves are absorbed by water vapor in the atmosphere.

A French graduate student, Charles Nordman, reasoned that if the atmosphere were blocking radio waves from space he had better put his own antenna up high somewhere to try to get above it. He carried it to the top of Mont Blanc. Nordman, too, failed to pick up radio waves from the sun—but in his case it was because of bad luck. His equipment would have worked at a time of solar maximum, when radio waves are emitted at peak levels. Unfortunately 1900 was a period of solar minimum and so he detected nothing. But Max Planck's work on black-body radiation and light quanta revealed another problem.

Hubble telescope photo of the constellation of Sagittarius, source of the radio signal detected by Jansky

Predicting from Planck's equations the amount of radiation received from the sun that should fall into the radio wave part of the spectrum (wavelength 4–40 in/10–100 cm), it became clear that the radiation would be very weak—too weak to be detected by equipment available at the time. A further blow came in 1902 when electrical engineers Oliver Heaviside (1850–1935) and Edwin Kennelly (1861–1939) predicted the existence of the ionosphere, a layer of ionized particles in the upper atmosphere that would reflect radio waves. (This layer has had important uses as an aid to radio communication, however. By bouncing radio waves off the ionosphere it is possible to transmit signals long distances.) These disappointing conclusions seem to have dampened enthusiasm for the search, and there were no more attempts to detect radio signals from space for 30 years.

The breakthrough came in 1932, when the American radio engineer Karl Jansky (1905–50) was employed by the Bell Telephone Company in New Jersey, USA, to investigate interference from radio static on its transatlantic telephone service. Using a large directional antenna, Jansky found a signal of unknown origin that repeated every 24 hours. He suspected it came from the sun, but then realized the repetition was actually every 23 hours and 56 minutes—less than a day-length. An astrophysicist friend, Albert Skellett, said it seemed to come from the stars. Using astronomical

Antenna of a radio telescope at the astronomy center of Yebes, Spain

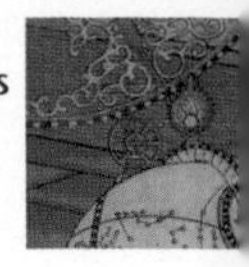

The remains of supernova SN1006 were produced by the explosion of a massive star some 7,000 years ago.

maps, they identified the Milky Way as the source, and more particularly the center of the galaxy, around the constellation of Sagittarius, as the peak of the signal coincided with the appearance of this consellation. Jansky suspected the signal came from an interstellar dust or gas cloud at the heart of the galaxy. He wanted to continue his work on radio waves from the Milky Way, but his employers moved him on to another project and he had to abandon his research. His one great discovery marked the start and end of his career in astronomy. Jansky's work inspired the American amateur astronomer Grote Reber (1911–2002), who built a parabolic radio telescope in his backyard in 1937 and carried out the first survey of the sky at radio frequencies.

Radio waves from the sun were first discovered in 1942 by James Hey (1909–2000), a research officer in the British Army. Radio astronomy was now becoming respectable: radio astronomers Martin Ryle (1918–84) and Antony Hewish (1924–) at Cambridge University mapped the sky's sources of radio in the early 1950s, producing the 2C and 3C surveys (Second and Third Cambridge Catalogues of Radio Sources).

Today, radio telescopes are often arranged in banks, their antennae pointing at the same area of sky and data pooled from all of them. Each telescope has a large collection dish that focuses the received radio waves on to the antenna. Using a technique called interferometry, developed by Ryle and Hewish, the data from each antenna are combined (or "interfered").

LITTLE GREEN MEN

The first name given to pulsars was LGMs, for Little Green Men, because of a suggestion that the pulses represented intentional radio transmissions by an alien life form. It caused such alarm that university authorities considered keeping the discovery secret. Then Jocelyn Bell discovered another pulsar, proving it was a natural phenomenon.

Coincident signals reinforce one another, while conflicting signals cancel each other out. The effect is to achieve the collecting power of a single gigantic dish. To minimize problems from the ionosphere and atmospheric water vapor, the best sites for radio telescopes are often located at high altitude in arid regions.

While radio telescopes can be used to investigate the sun and planets of the solar system, they have been most useful in exploring objects so distant they cannot be seen at all using optical telescopes. This has led to major discoveries such as quasars and pulsars.

Quasars—Mighty and Remote

Quasar is short for "quasistellar object." Quasars are very energetic objects with a very large red shift (see page 191), which means they are extremely remote. There are 200,000 known quasars, all between 780 million and 28 billion light years away, which makes them the most distant objects of which we have any knowledge. The first quasars were spotted in the late 1950s, and described by the Dutch astronomer Maarten Schmidt (1929–) in 1962. Massive bursts of radiation from quasars may be produced by the release of gravitational energy as matter falls toward a massive Black Hole. Up to 10 percent of this mass is converted to energy that is able to escape before the event horizon (see page 187). Nuclear fusion operating within stars could not produce enough energy to make a quasar glow brightly enough (with visible light and other forms of electromagnetic radiation) to be detected from Earth at such vast distances. The explosion of a supernova could produce enough energy to be seen for a few weeks, but a quasar persists. For the most distant quasars to be visible, they must be two trillion (2×10^{12}) times as bright as the sun. Or were—these objects are billions of light years away, so we are seeing them as they were near the beginning of the universe.

Up, Up, and Away

Our understanding of astronomy and the physics of space changed considerably throughout the 20th century. But perhaps the most important development was the marrying of time and space into a single concept—the space-time continuum, discussed in the next chapter.

Maarten Schmidt

PULSARS—SPINNING BEAMS OF POWER

A pulsar is a highly magnetized, rotating stellar body. It forms when a massive star's fuel resources become depleted and its core collapses down to an incredibly dense body called a neutron star. The pulsar is so-called because as it spins it emits highly directional radiation that can only be observed when it points directly at Earth—creating a pulse rather like the beam of a lighthouse flashing across the sea. The intervals between pulses range from 1.4 milliseconds to 8.5 seconds. The rate slows down until it eventually turns off after a period of 10–100 million years, so most pulsars that ever formed (99 percent) are no longer pulsing.

Jocelyn Bell Burnell

The first pulsar was discovered in 1967 by a 24-year-old PhD student Jocelyn Bell (now Dame Jocelyn Bell Burnell). Controversially, it was her supervisor Antony Hewish who was awarded the Nobel Prize (in 1974) for the discovery, and not her. Observations in 1974 of a pulsar in a binary system (in which a pulsar orbits a neutron star, with an orbital period of eight hours) provided the first evidence of gravity waves, confirming another part of Einstein's general relativity theory.

As a pulsar spins, its radioactive emissions can only be detected from Earth in pulses.

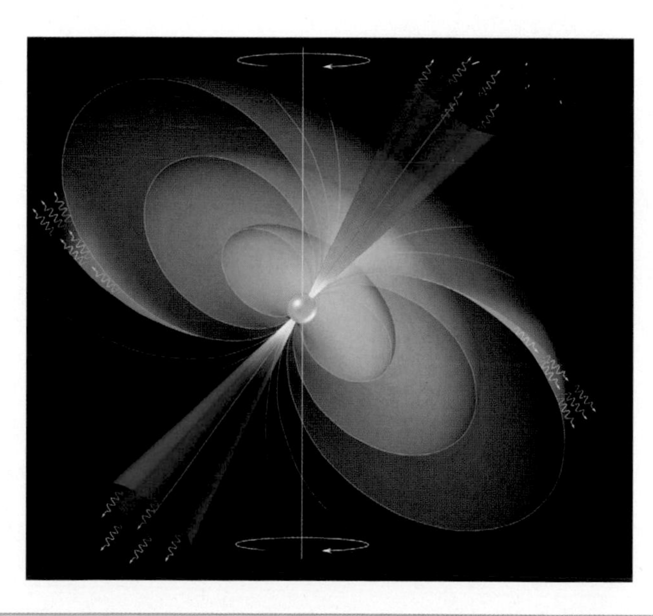

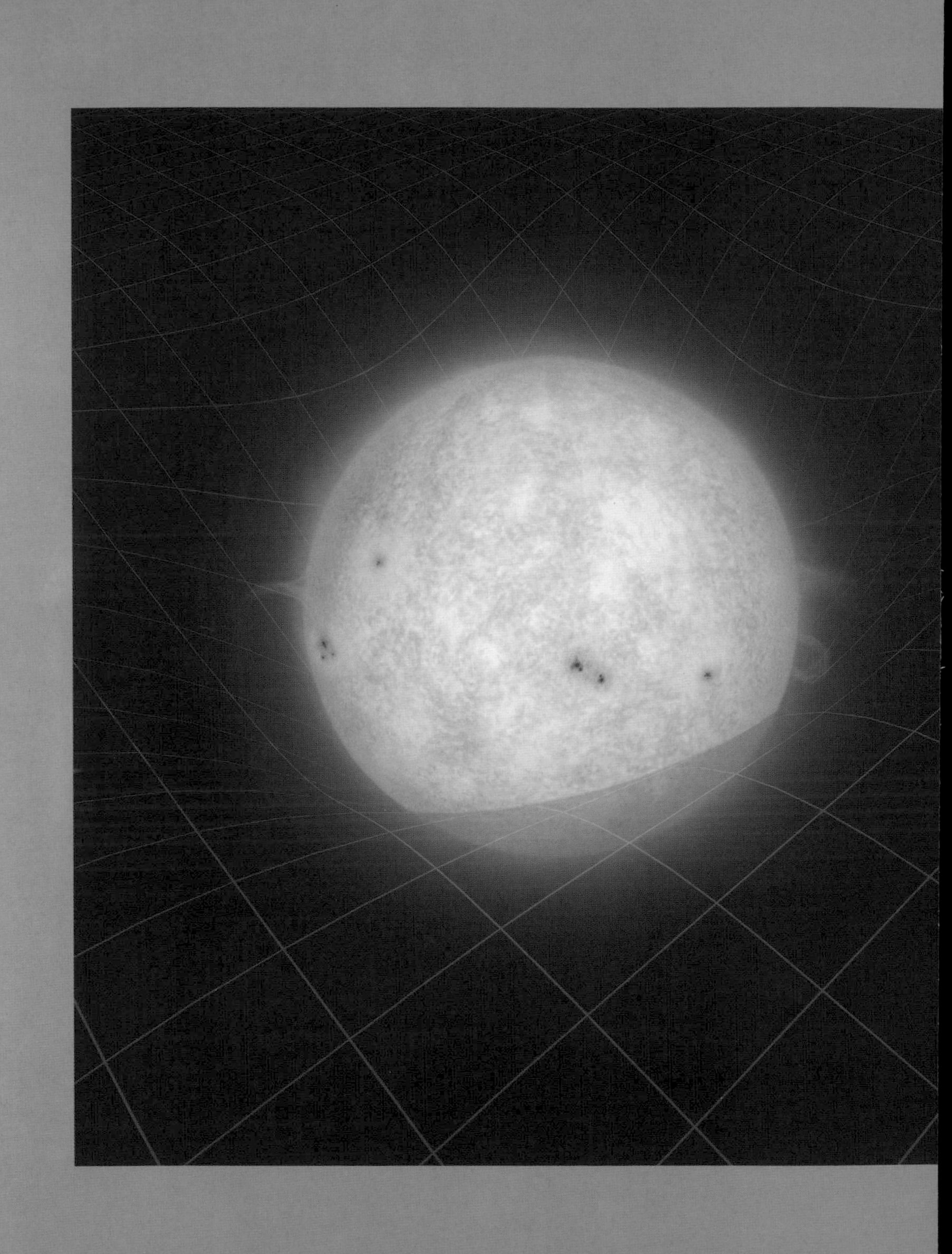

CHAPTER 7

SPACE-TIME Continuing

For thousands of years, staring into space and wondering about its strange geography was just that—it was looking outward, trying to see how the stars and planets, the sun and moon, relate to the earth. The movements of the sun and moon were humankind's celestial clock, measuring the hours, days, months, and years. But space and time were regarded as separate concepts. From the beginning of the 20th century, though, our relationship with space and time began to change. After Einstein, they became locked together as the space-time continuum, and the study of space became focused not just on "what's out there" but on the past and possible future of our universe.

A star distorting the space-time continuum, creating a gravitational effect

A Brief History of Time

Although it's easy to see the passage of days, the pattern of a whole year becomes evident only with recording and counting. The earliest evidence of people keeping track of time dates from around 20,000 years ago. Mathematics and early astronomical knowledge probably emerged together as people learned to track and predict the movements of the heavenly bodies.

A clepsydra was used to measure time in Ancient Greece; water clocks have been in use for thousands of years.

Measuring the course of a day was achieved early on using a gnomon, an object such as the pointer on a sundial that projects a shadow to track the progress of the sun across the sky. For millennia this remained the best guide to the passage of time. Then, in the 17th century, Galileo compared a swinging lamp with his own pulse and discovered the regular movement of a pendulum. The pendulum always takes the same time to swing: as the arc decreases, so the pendulum's movement slows down to keep the interval regular.

Galileo designed a pendulum clock, but never made one. It was Christiaan Huygens who built the first pendulum clock in 1656. Later, Robert Hooke used the natural oscillation of a spring to control the mechanism of a clock. The measurement of time by mechanical means remained the norm until 1927 when Canadian-born telecoms engineer Warren Marrison, working at Bell Telephone Laboratories in New Jersey, discovered that he could measure time accurately using the vibrations of a quartz crystal in an electric circuit.

> *"My soul yearns to know this most entangled enigma. I confess to Thee, O Lord, that I am as yet ignorant what time is."*
>
> St Augustine

Tomorrow and Tomorrow and Tomorrow

Clocks measure linear time, which is very convenient for human lives, but may not represent the whole story. The idea that time might not be linear was suggested by both Buddha and Pythagoras around 500 BCE. They believed that time could be cyclic and that a human being, after dying, may be reborn. Plato thought time was created at the beginning of all things. But for Aristotle, time only existed where there was motion. An apparent paradox, proposed by the philospher Zeno (*ca*.490–430 BCE), seems to show that neither time nor motion can exist. If we divide time into ever smaller portions, the distance traveled by a moving arrow becomes ever shorter until, in the instant of "now," the arrow does not move. But in that case it cannot exist or move at all, as time is made up of an infinite number of "nows" in which no movement is

The clockwork mechanism provided the first way of telling the time accurately.

> *"... absolute, true, mathematical time... from its own nature, flows equably without relation to anything external."*
> Isaac Newton

St Augustine

taking place. The Christian philosopher St Augustine of Hippo (354–430 CE) came to the conclusion that time did not exist unless there was an observing intelligence as it was only the remembrance of things past and the expectation of future events that gave time any existence outside the present.

The French mathematician Nicole Oresme (1323–82) asked whether celestial time—time measured by the movement of the heavenly bodies—was commensurate: that is, whether there was a unit in which their movements could all be measured in whole numbers. He suggested that an intelligent creator would surely have made them so, but stopped short of finding that the lack of a common measure means there is no God.

Marrying Space and Time

Our personal experience of time is straightforward. Time moves from the past through the present toward the future, with no chance to go back, jump forward, or freeze-frame. It moves at a steady rate in one direction. It's hardly surprising that for millennia we assumed this was the very nature of time. But maybe it isn't.

Everything is Relative

All movement is relative to the position or movement of the observer. So you may walk across the room, and someone standing still in the room will judge your speed to be around 3 miles (5 km) per hour. Both you and the observer are actually on a spinning globe whirling through space at nearly 19 miles (30 km) per second, but only your movement across the room is noticeable. An observer on a distant planet (with a good telescope), though, would also see the spinning and whirling globe. (Galileo realized this, although he talked about a person on a ship, viewed by a spectator on the shore rather than an alien with a telescope.) So the speed at which an object moves depends on the frame of reference; movement can only be measured relative to other objects or observers. The frame of reference may be the same room, the same ship, the same planet, or the same galaxy.

Einstein found an exception to this basic rule: light, he said, always travels at the same speed—regardless of the speed at which an observer is moving. So, no matter how fast you are traveling, a beam of light will whizz away from you at 186,282 miles (299,792,458 meters) per second. As the

TAKING GRAVITY TO EXTREMES: BLACK HOLES

Black Holes are "singularities" in space-time. They are areas where gravity is so powerful that not even light can escape, and anything passing too near gets sucked in. Black Holes may form when stars collapse in on themselves to become very tiny, in some cases no larger than the nucleus of an atom, and exceedingly dense. The escape velocity required to leave a Black Hole is greater than the speed of light. The size of a Black Hole is measured by its event-horizon—the boundary over which nothing can escape. Although an astronaut falling into a Black Hole may not notice anything unusual as he crosses the event-horizon, an observer viewing from outside will see time for that person slow down. Poised on the brink of the event-horizon, they appear to be frozen in time.

The concept of Black Holes (although not the name) was first suggested by two people independently—Pierre-Simon Laplace (1749–1827) in 1795 and before him the English philosopher John Michell (1724–93) in 1784.

Michell called the phenomenon of a star so dense and with such a fierce gravitational pull that light could not escape it a "dark star." The idea was revived by the German physicist Karl Schwarzschild (1873–1916) shortly before his death in 1916 when he calculated the gravitational fields of stars and collapsed stars. The term "Black Hole" was coined by the American theoretical physicist John Archibald Wheeler (1911–2008) in 1967 when cosmologists found the first evidence for their existence.

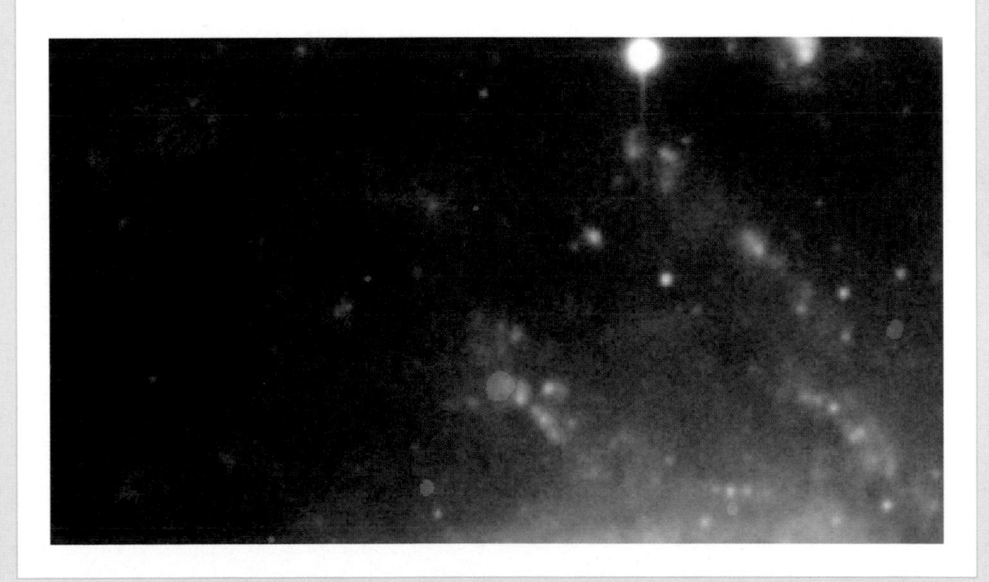

Hubble telescope image of a supernova, the bright spot low down to the left

speed of light is constant, other things cannot be—and one of those things is time. In fact, approaching the speed of light, time slows down and distance contracts. Einstein was proved to be correct in this regard in 1971. An atomic clock taken for a trip on a very fast plane recorded a slightly shorter time than an identical clock left stationary on the ground. Traveling in a fast plane is not a good way of extending your life, though—you would need to make 180 billion circuits of the Earth to save a single second.

Einstein's theory of general relativity, published in 1915, went further, bringing together space-time and matter and using gravity to explain the effect of one on the other. Matter bends space-time, rather as a ball thrown on to a stretched blanket makes a dip in the blanket. The way other objects, and light, move as a response to this bending we describe as gravity. So, just as a small ball will naturally roll toward the dip in a blanket created by a large ball, a small body in space will naturally gravitate toward a larger one, constrained by the curvature of space-time. This curvature had been proposed long before Einstein by the German mathematician Bernhard Reimann (1826–66), whose ideas were published after his death, in 1867–8. But Einstein went much further than Reimann in that he provided equations to explain and predict the curvature.

Far Away and Long Ago

There is another, less theoretical and complex way in which our interest in space has become tangled up with time and the speed of light. When we look at the stars, we are seeing them as they were in the past because of the length of time it takes for their light to reach us. Even the light from the sun is eight minutes old by the time we see it. If the sun had been turned off two minutes ago, we would continue to see it shine, oblivious of the impending disaster, for another six minutes.

The light from the nearest star, Proxima Centauri, takes four years and three months to reach us. One of the brightest stars ever detected, first spotted in 1988, was a supernova. Since a supernova represents the death of a star, one that has exploded, that star no longer exists. It was five billion light years away, so the light seen in 1988 signified the star's death five billion years ago, before our own solar system had even formed. The supernova witnessed by Kepler and Galileo in 1604 is around 20,000 light years away—so that star ceased to exist at around the time mammoths roamed ice-bound Europe.

Back to the Beginning

Of course, when no-one knew what the stars and planets were, it was difficult to say how they got to be there, and with a few notable exceptions most cultures left this question to religion. Archbishop James Ussher (1581–1656) calculated the date of creation (from which the age of the universe could be estimated), to be October 22, 4004 BCE, based on the genealogies recorded in the Bible. Many other societies have proposed their own creation dates. The Mayans gave a date for creation that translates as August 11, 3114 BCE. Judaism placed creation on either September 22 or March 29, 3760 BCE. The Puranic Hindu religion went in the other direction, with an extravagant date for creation of 158.7 trillion years ago. There are also suggestions that the universe has always been there. Aristotle, for instance, thought the universe finite but eternal.

Out of Chaos

Anaxagoras, in the fifth century BCE, suggested that the universe began as a pile of undifferentiated, inert matter. At some point, after an infinity in which nothing happened, Mind (his analogy for the natural laws of the universe) began to act on this matter and set up a whirling motion. As a consequence, denser matter clumped together and less dense matter drifted to the outside of the bodies so formed, or wafted about between them. This is not so very different from the model that modern astronomers have of the development of the universe, with solar systems forming as pre-planetary disks coalesced from a vast cloud of dust and, through the action of gravity and centripetal force, formed into planets. Anaxagoras worked only from logic (and a good deal of imagination).

The philosophers Democritus and Leucippus (fifth century BCE) believed the cosmos formed when whirling motion led atoms to clump together into matter. As the universe is infinite in time and space and contains an infinite quantity of atoms, all possible worlds and configurations of atoms will exist, and so the existence of our world and of humanity is not special, but rather inevitable. As everything is in constant flux, a cosmos will come into being and eventually disintegrate and its indestructible

"[Mind ruled] this rotation in which now revolve the stars and sun and the moon and the separated air and the aither. And the dense separates from the light, the hot from the cold and the bright from the dark and the dry from the wet." Anaxagoras, fragment B12

Descartes' division of space into regions containing particles rotating around a center, 1644

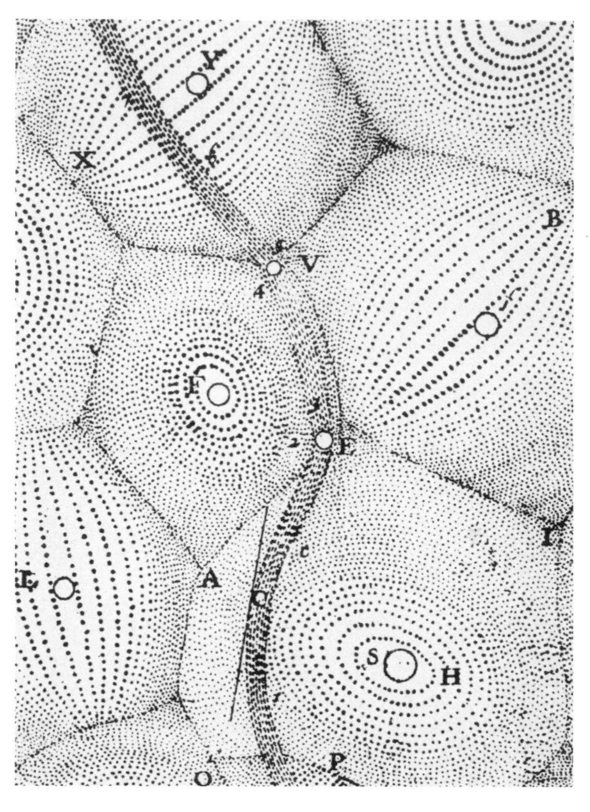

atoms will be reused in a new cosmos. Even on a shorter time span, we know that the atoms in a star system that dies are eventually recycled.

René Descartes described a "vortex" universe in which space was not empty but filled with matter that swirled around in whirlpools or vortices producing what would later be called gravitational effects. In 1687, Newton proposed a static, infinite, steady-state universe in which matter is evenly distributed (on the large scale). His universe was gravitationally balanced, but unstable. It endured as a scientific model until the 20th century. Even Einstein accepted it as a given truth until discoveries proved otherwise.

The Modern Universe

One feature of Einstein's general relativity equations is that they cannot work in a static universe without a "fudge." As Einstein firmly believed that the universe was static, he added a "cosmological constant" to his equations to make them work. But others interpreted his equations differently. An expanding universe was first proposed by the Russian cosmologist and mathematician Alexander Friedmann (1888–1926). Using Einstein's relativity equations, Friedmann put forward a mathematical model of an expanding universe in a paper published in 1922. He died the following year at the age of only 37 of typhoid, a disease

The Greek Stoic philosophers in the third century BCE believed the universe was like an island surrounded by an infinite void and was in a constant state of flux. The Stoic universe pulsates, changing size and suffering periodic upheavals and conflagrations. All parts are interconnected, so that what happens in one place affects what happens elsewhere, an idea curiously mirrored in quantum entanglement (see page 125).

RED SHIFT

If the light from a star is analyzed using spectroscopy, its spectrum will be seen to be "squeezed" toward blue wavelengths if moving toward the observer (blue shift) and "stretched" toward red wavelengths if moving away (red shift). This is called the Doppler effect. A similar effect occurs with sound waves; a police car siren will have a higher pitch when it approaches the listener, as its sound waves are compressed, and a lower pitch when it moves away, as its sound waves are stretched. The red shift that Hubble observed, however, is not the result of a Doppler effect caused by the movement of the galaxies' stars (although this would cause a red shift). Instead, it is a result of the space between our galaxy and distant galaxies stretching, which is how the universe expands. The wavelength of light traveling through that stretching space is also pulled out and extended. Light with a longer wavelength is redder, hence the red shift. This is why the existence of

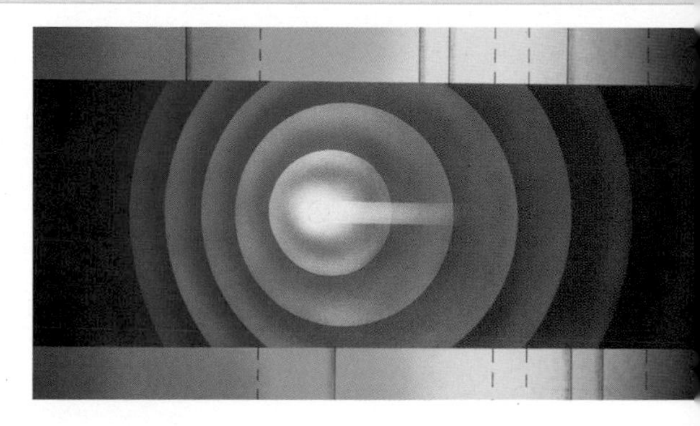

Light waves are shifted toward the red or blue end of the spectrum depending on whether the source is moving toward or away from the observer.

the red shift is evidence for an expanding universe. The red shift of some distant galaxies was first measured by the American astronomer Vesto Slipher (1875–1969) and described in 1917. But it was Hubble who discovered that the red shift was universal and that the farthest galaxies were receding fastest. He published this as the "Relation between distance and radial velocity among extra-galactic nebulae."

he contracted while on holiday in the Crimea, and his work was largely overlooked. Einstein was one of the few to read Friedmann's paper, but he rejected it out of hand. However, Einstein was forced to reject his own earlier model and drop the cosmological constant after evidence emerged that Friedmann had been right.

The American astronomer Edwin Hubble (1889–1953) demonstrated in 1929 that distant galaxies were moving away from our region of space in all directions. Hubble had analyzed these galaxies spectroscopically and noticed that their spectra were shifted toward the red end of the spectrum—a so-called "red shift" (see box, above). These findings were taken as evidence that the universe is indeed expanding. Einstein now broadly followed Friedmann's model but took the view that the universe oscillates between expansion, following the Big Bang, and then contraction, as eventually gravity

pulls all matter in again, resulting in a Big Crunch and a singularity, which will explode in another Big Bang. The cycle continues forever, but as time is one with space, both space and time are without beginning or end (or have infinite beginnings and endings, depending on how you wish to look at it).

From Cosmic Egg to Big Bang

The modern view of the universe came into being with the theories of the Belgian priest and physicist Georges Lemaître (1894–1966). Lemaître expressed the view that the universe started as an infinitely small and infinitely dense point—now called a singularity, but called by Lemaître a

GEORGE GAMOW (1904–68)

George Gamow was born in Odessa, in the Russian Empire, an area that is now part of Ukraine. Gamow was a hugely successful and versatile physicist who made a number of important discoveries and hypotheses. His parents were both teachers, though his mother died when Gamow was only nine. His education was interrupted by the destruction of his school by shelling during the First World War, and as a consequence he was largely self-taught. Gamow worked with some of the greatest European physicists of his day, including Rutherford and Bohr. He made two attempts to escape the USSR, the first by kayaking 155 miles (250 km) across the Black Sea to Turkey, and the second by traveling from Murmansk to Norway. Both attempts were foiled by bad weather. Gamow eventually defected, along with his wife, while attending the Solvay Physics Conference in Belgium in 1933 and settled in the USA in 1934.

Gamow's work spanned quantum mechanics and astronomy; he developed the "liquid drop" model of the atom, in which the nucleus is considered a drop of incompressible nuclear fluid, described the inside of red giant stars, worked out alpha particle decay and explained that the reason 99 percent of the universe comprises hydrogen and helium is because of reactions that took place in the Big Bang. He predicted the existence of cosmic microwave background radiation, theorizing that the afterglow of the Big Bang would still persist after billions of years. His estimate was that it would now have cooled to about five degrees above absolute zero. When Penzias and Wilson discovered CMBR in 1965 (see box, opposite), they found the temperature is actually 2.7 degrees above absolute zero.

A NOBEL PRIZE BY ACCIDENT

In 1978, Arno Penzias and Robert Wilson shared the Nobel Prize in Physics for their discovery of cosmic microwave background radiation. In fact, they had not been looking for it, and did not at first recognize it when they found it. Penzias and Wilson were tuning a sensitive microwave antenna at Bell Telephone Laboratories in Holmdel, New Jersey, for use in radio astronomy when they picked up interference that was disrupting their work. They could not get rid of it. It was constant and came from all parts of the sky equally. In fact they had stumbled on cosmic microwave background radiation (CMBR). Not far away, at Princeton University, the team of Robert Dicke, Jim Peebles, and David Wilkinson were building equipment specifically to look for CMBR and quickly realized what Penzias and Wilson had found. On hearing the news, Dicke turned to the others and said, "Boys, we've been scooped."

primeval atom or "cosmic egg." An unimaginably powerful event that we now call the Big Bang exploded this singularity, transforming all the matter of the universe and blasting it through space.

Lemaître presented his idea of an expanding universe at the Solvay Physics Conference in Belgium in 1927 when he gave the first statement of what would later become Hubble's Law—that the speed of distant objects moving away from the Earth is proportional to their distance from Earth. Lemaître discussed this with Einstein at the conference, but Einstein again rejected the theory. He told Lemaître, "Your maths is correct but your physics is abominable!" However, Hubble's discovery confirmed Lemaître's physics, demonstrating that the red shift in light from faraway galaxies is proportional to their distance from Earth.

Despite his success, Lemaître's theory of the "cosmic egg" was derided, even by Eddington, who had championed his model of the expanding universe. The name Big Bang originated with a sarcastic remark by the British astronomer Fred Hoyle (1915–2001) in 1949. Hoyle continued to favor a "steady state" model of the universe, long after the general consensus was that Lemaître was right. Although Hoyle's universe, described in 1948, expanded, it included the regular insertion of new material to keep the overall density stable. The main argument against the Big Bang theory was that there should be some heat energy left over from the original event, which should be detectable. The physicist George Gamow (see box, opposite) had theorized that with the expansion of the universe, this heat energy would have cooled, shifting into the microwave band. Confirmation came in 1965 with the accidental discovery of cosmic microwave background radiation (CMBR) by two radio astronomers, Arno Penzias and Robert Wilson in 1965 (see box, above). With this evidence, most of the remaining dissenters moved over to the Big Bang camp.

How Many Stars?

The earliest star catalogs could list only those stars visible to the naked eye. As technology improved, first with the ocular telescope and then with radio

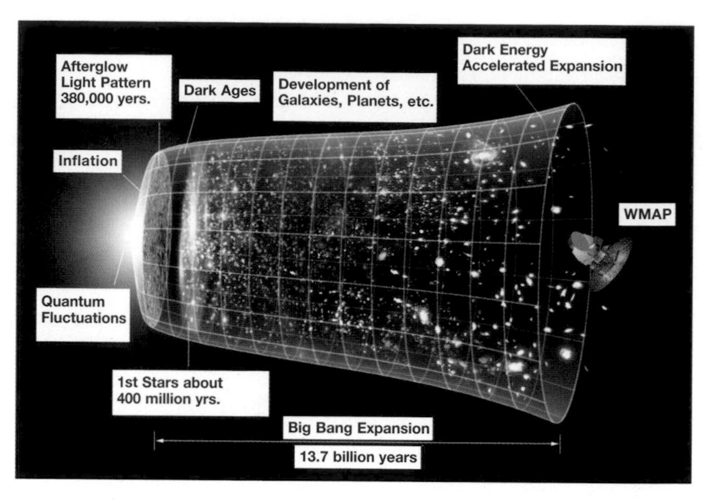

How the universe has evolved since the Big Bang

telescopes, the number of detectable stars multiplied—steadily and then exponentially. The Draper star catalog (see page 170) eventually listed 359,083 stars. Yet the estimated number of stars in the universe far exceeds any catalog and, just like the universe, tends to expand. Until late 2010, the generally accepted estimate was between 10^{22} and 10^{24} stars. Then a research team led by Pieter van Dokkum at the Keck Observatory in Hawaii discovered in 2010 that there may be three times as many stars as previously thought, on account of a proliferation of previously invisible red dwarf stars (perhaps 20 times more than previous estimates in some galaxies).

The Observable Universe

We now have various ways of estimating the age of the universe: by measuring the abundance of radioactive isotopes such as uranium-238 and their decay products (nucleocosmochronology); by measuring the rate of the universe's expansion and calculating backward to work out when it must have started; and by looking at globular clusters of stars and deducing their age from the types of star they contain. The most accurate figure for the age of the universe is currently thought to be 13.7 billion years. It is based on data from NASA's Wilkinson Microwave Anisotropy Probe, a spacecraft that measures the cosmic microwave background radiation.

That the most distant known quasar is around 28 billion light years away (see page

The explosion of a supernova shown in optical (left), ultraviolet (center), and X-ray (right) wavelengths

180) may seem impossible if the universe is only around 13.7 billion years old. The anomaly is accounted for by the expansion of space-time between Earth and the quasar. The light we now receive from the quasar was emitted perhaps 12.7 billion light years ago when the quasar was closer to Earth, but as the space between the two has since increased, the quasar is now much farther away. Although neither light nor a body can travel through space at speeds greater than the speed of light, space-time can expand at any rate. It is thought that the observable universe (that which could theoretically be observed if we had the right technology) is around 93 billion light years across. That does not put a limit on the size of the whole universe. Beyond that, there may be matter which is now separated from Earth by so much intervening space that its light has still not reached us.

BIG BANGS

Until 2010 there was no evidence to suggest the Big Bang may have been one in a cycle of expanding and contracting universes, but then Sir Roger Penrose (1931–) and Vahe Gurzadyan (1955–) discovered clear concentric circles within the microwave background radiation, which suggests regions of the radiation have much smaller temperature ranges than elsewhere. This, they argue, suggests a previous, older Big Bang, preserved as a type of fossil in the CMBR.

How Many Universes?

Although the word "universe" means there is only one, a few scientists have suggested there is actually a multiverse in which our own universe is only one of many. The theoretical physicists Hugh Everett III (1930–82) and Bryce DeWitt (1923–2004) suggested a "many worlds" model in the 1960s and 1970s, and the Russian-American physicist Andrei Linde (1948–) described in 1983 a model in which our universe is one of many "bubbles" formed in a multiverse subject to eternal inflation.

All Downhill From Here

Our own sun is around half way through its likely life. It can be expected to last a few billion years yet before it follows the pattern observed elsewhere in the universe of expanding into a red giant, then collapsing into a white dwarf and finally growing cold.

Although we clearly won't be around to witness it, the end of the universe—if there ever is one—concerns some cosmologists. Will it expand forever, until it is a thinly dissipated soup of matter, no longer cohering into useful planetary systems? Or will it all be sucked back into a Big Crunch, ready to burst forth again in a new Big Bang? If so, this cycle may be eternal (though the word has no meaning in a system in which time, along with space, is crushed to nothing and recreated to start from scratch). The beginning and end of the universe are truly the frontiers of science, areas we explore with logic and mathematics—but even here there are experimental methods that will help to refine our theories as we mold the physics of the future.

RIK NITSCHE

CHAPTER 8

PHYSICS for the Future

When Max Planck said in 1874 that he wanted to specialize in physics, his tutor advised him to pick a different subject as there was nothing left to discover in the physical sciences. Luckily, Planck ignored him. There is still, nearly 150 years later, much left to be discovered in physics. We cannot reconcile gravity and quantum mechanics; we cannot account for most of the mass of the universe; there are particles we can't detect but suspect are there, somewhere, to be found; we can't quite explain what energy is and we don't know what the fate of our universe will be, or whether it is unique or just one among many. These are some of the questions ready to be addressed by the physicists of the future, who are still in our schoolrooms and university lecture halls.

Practical applications of physics harness the natural laws of the universe for new technology.

Ripping It Up and Starting Again

Physics in the 20th century led to a fundamental reappraisal of much of what had gone before, combining space and time into the space-time continuum, replacing certainties with uncertainty and probabilities, turning particles and waves into wave-particle dualities, and introducing other ideas that, although bizarre, could not be rejected. In fact, the new theories did not so much overturn what had gone before as encompass them in something larger. The something larger, though, still does not account for everything and ultimately it, too, must be incorporated into a set of theories or models that accounts for everything we have so far discovered, as well as what is still unexplained.

Is That All?

It seems a bit of a failure for physics, but one of the biggest remaining problems is how to account for 96 percent of the mass-energy density of the universe. The universe that we can see, because it either reflects or emits light, only accounts for a tiny amount of what is known to be there, around 4 percent. The term "dark matter" was coined to describe matter that we know is there but cannot see. The idea of dark matter was first proposed by the Bulgarian-Swiss astronomer Fritz Zwicky (1898–1974) in 1933.

Zwicky applied calculations derived from Einstein's relativity theories to gravitational interactions observed in the Coma galaxy cluster and found that the cluster must contain hundreds of times more mass than its overall luminosity would suggest. He proposed that the balance was made up of dark matter.

So what is this mysterious substance? The most widely accepted current theory divides dark matter into baryonic and nonbaryonic matter. Baryonic matter is ordinary matter made up of protons and neutrons and the like. All the visible objects in the universe must emit or reflect light. That might seem fairly obvious, but it is

This ring of dark matter formed by the collision of two galaxies was photographed by the Hubble telescope in 2004.

very significant. If a planet wanders where it is not illuminated by any star, or if a star burns out, it can no longer be seen. Dark baryonic matter is likely to be made up of invisible matter such as gas clouds, exhausted stars, and un-illuminated planets. These are called MAssive Compact Halo Objects (MACHOs). The presence of MACHOs can be inferred from the gravitational effects they have; they were first found in the Milky Way in 2000.

There are not enough MACHOs to supply all the dark matter, though. The vast majority of dark matter is thought to comprise Weakly Interactive Massive Particles (WIMPs). These particles are, by definition, difficult to find as they don't interact with other matter through electromagnetic forces. Some dark matter may be accounted for by neutrinos (see page 135), but there is still space for other undiscovered and theoretical particles such as axions and even more untheorized exotic particles.

ARTIFICIAL METEORS

Fritz Zwicky had a novel and unconventional approach to astronomy and many of his ideas (including dark matter) were not taken seriously by his contemporaries. In October 1957, Zwicky fired metal pellets from the nose cone of the Aerobee rocket, making artificial meteors that were visible from the Mount Palomar observatory. One of the pellets is thought to have escaped the gravitational field of the Earth and become the first man-made object to go into solar orbit.

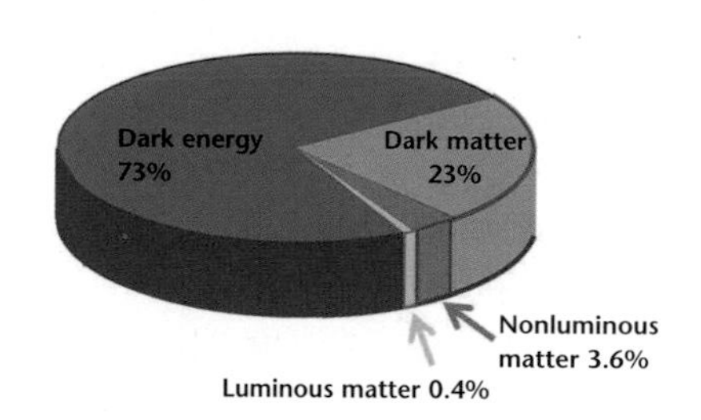

DARK ENERGY

If the existence of dark matter was hard to accept, cosmologists had a greater shock in store when the results of the Supernova Cosmology Project were announced in 1999. This study had looked at Type 1a supernovae, a type of exploding star whose mass and luminosity is known and therefore whose red shift (see page 191) can be calculated accurately. The project's findings revealed that the universe was not expanding at a steady rate or slowing down, as had been assumed, but accelerating. This acceleration has since been confirmed by other research, including detailed studies of the CMBR. To account for this phenomenon, scientists coined a new term—dark energy.

Even with MACHOs and WIMPs, the universe's mass-energy budget is massively in deficit. It is now estimated that nearly three-quarters (around 74 percent) of the mass-energy of the universe is accounted for by mysterious dark energy, with dark matter accounting for most of the rest. Dark energy is thought to have strong negative pressure and so to account for the accelerating expansion of the universe. It is probably homogenous, not very dense, but present everywhere that is otherwise considered empty space. One contender for the dark energy title is the cosmological

> *"The universe is made mostly of dark matter and dark energy, and we don't know what either of them is."*
>
> Saul Perlmutter, of the Supernova Cosmology Project, 1999

constant, originally added by Einstein as a fudge in the general relativity equations to explain why the universe was not collapsing under the force of gravity. Einstein later abandoned the idea, but it is now being resuscitated to explain these new findings.

One theory is that the cosmological constant acts like anti-gravity, preventing gravity from pulling the universe in on itself. The force of the cosmological constant at the moment is thought to be a little larger than the force of gravity, but it is not known whether it has always been the same or will always be the same or whether it is truly a constant. Not all cosmologists accept the idea of a cosmological constant and have put forward other, even more esoteric ideas such as "string theory" (see page 202). No convincing evidence has yet been found to make any particular theory seem overwhelmingly likely.

Where Next for Matter?

The standard model of matter is that atoms are composed of composite particles such as neutrons and protons and that these are composed of elementary particles such as quarks (see page 133). A whole host of other particles are theorized but not yet proven to exist—or may no longer exist. Exploring

How the universe has looked since the Big Bang

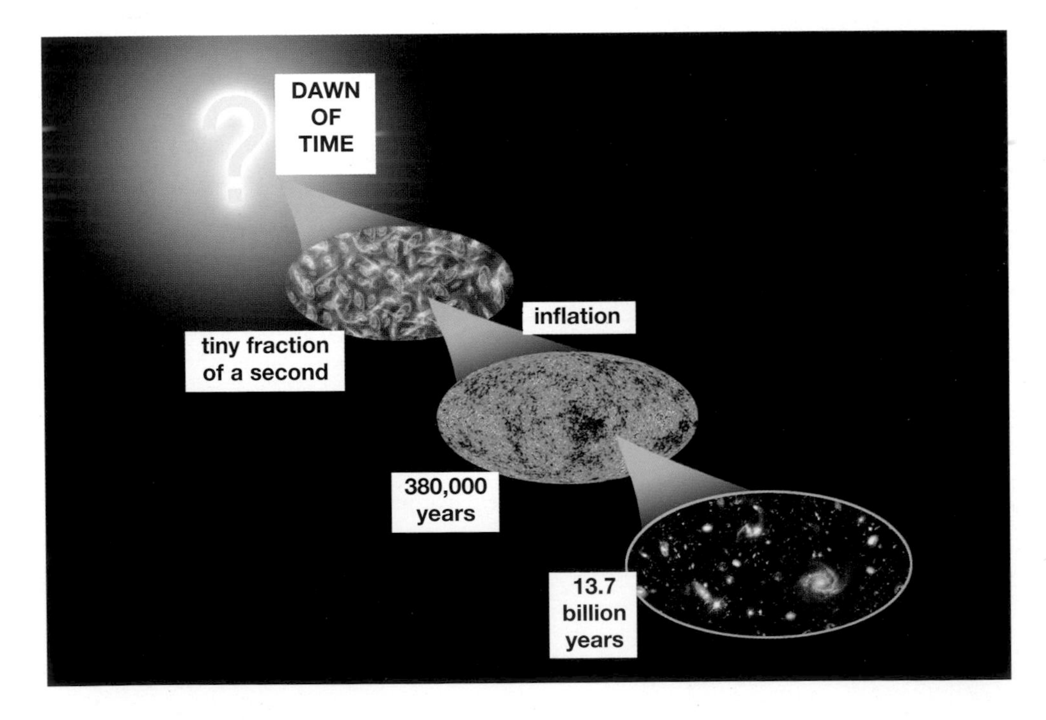

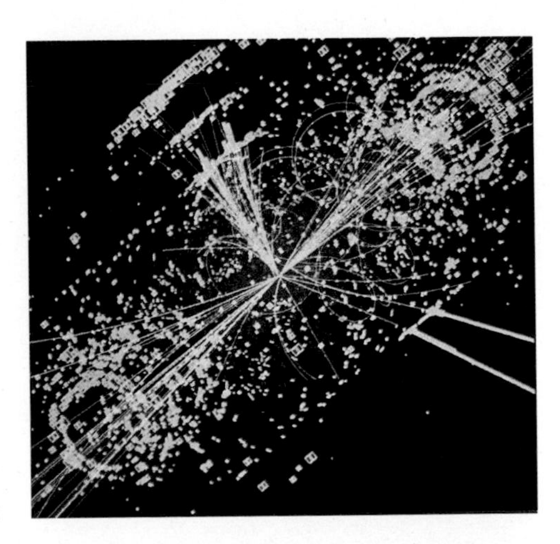

Simulation of the creation and decay of a Higgs boson; it produces two jets of hadrons and two electrons.

these particles experimentally rather than as a mathematical model is complex and costly, requiring highly sophisticated equipment, especially as many have a very short lifespan.

The postulated Higgs boson (or "God" particle) is the only elementary particle predicted by the standard model of matter that has not yet been detected. It is thought to impart mass to matter, and was first suggested by the English theoretical physicist Peter Higgs (1929–) in 1964.

To understand this, it's necessary to look for a moment at the particles that mediate the four fundamental forces: electromagnetism is mediated by virtually massless photons; gluons link quarks by the strong nuclear force; and W and Z bosons carry the weak nuclear force and are very heavy, relatively speaking—about 100 times the mass of a photon. The problem for physicists is to account for the difference in mass of these force-mediating particles. The solution is a model that has some of the particles effectively wading through treacle.

The Higgs field is something like a force field through which matter has to move in space. Some quantum particles are slowed more than others as they move through it. The slowing of a particle effectively imparts mass to it. Photons are not inhibited by the field and have very little mass, but W and Z bosons are slowed considerably by the field and so have significant mass. The Higgs field is mediated by the Higgs boson. If the Higgs boson could be proven to exist, the standard model would be complete.

How do we look for such a particle, though? Physicists are currently trying to blast one into visibility using huge particle accelerators such as CERN's Large Hadron Collider (LHC) in a tunnel under Geneva, and at Fermilab's Tevatron near Chicago. The existence of the "top" quark was confirmed at Fermilab in 1995. These accelerators fire beams of particles at extremely high speed in opposite directions around a circle so that they collide. The LHC is the largest such machine, with a circular tunnel around 17 miles (27 km) in circumference. The LHC fires beams of protons 11 months of the year and lead ions one month of the year.

The beams of protons are accelerated to within 10 feet (3 m) per second of the speed of light and set off in bursts so that collisions do not happen continuously but always at least 25 nanoseconds apart. It takes an accelerated proton only 90 microseconds to complete one circuit of the collider tunnel—equivalent to 11,000 circuits per second. The research program at the LHC began in 2010. Physicists expect

Stephen Hawking in zero gravity aboard a modified Boeing 727 aircraft

that, if the standard model is correct, a Higgs boson will be produced every few hours; it will take two or three years' data to confirm that this has happened.

Separating Baby from Bath Water?

Einstein struggled—and failed—to find a unifying theory that would explain everything, bringing gravity and quantum mechanics together in a comprehensive set of equations. Anaxagoras could have said the same. He wanted a single explanation for the movement and change of state that could account for all the change that occurs in the physical world. He insisted that it should not have any superstitious or divine component and must be entirely logical. The cosmic Mind, in his model, constantly surveyed, regulated, and managed the infinite changes taking place to make sure they were all in order. What he meant was that there was a law, which he had not discovered or explained, that controlled the flow of all matter. This was an unsatisfactory explanation, as his successors noted, but one that is not very different from the beliefs of Einstein and Hawking that there must be a unifying theory, if only we could discover it. At the end of his life, Einstein acknowledged that he was not going to succeed and must leave that work for others. It has still not been achieved, and the gulf between quantum theory and general relativity—despite the experimental evidence that both are correct—remains a major puzzle for physicists.

One approach to this problem has been the development of string theory. It is not yet a single coherent theory, cannot be tested, and may not be widely accepted, but it endeavors to unite quantum theory and general relativity by providing a deeper description of both. In string theory, all subatomic particles are tiny fragments of

> *"M-theory is the unified theory Einstein was hoping to find... If the theory is confirmed by observation, it will be the successful conclusion of a search going back more than 3,000 years. We will have found the grand design."*
> Stephen Hawking, *The Grand Design*, 2010

"string," either open-ended or looped, that vibrate in many dimensions. The difference between particles comes not from their composition, which is all the same, but from the harmonics of their vibrations. And these vibrations take place not just in the three dimensions of space and one of time with which we are familiar, but in 10 dimensions. Some of these may be curled in on themselves, or last only a very short time, so that we are not aware of them. String theory is highly speculative, and even its proponents have very different versions of it.

M-theory is a development of string theory that takes theoretical physics to new frontiers. The addition of an eleventh dimension is its most modest contribution. To vibrating strings, it adds point particles, two-dimensional membranes, three-dimensional shapes, and entities in more dimensions that are impossible to visualize (p-branes, where *p* is a number in the range zero to nine). The way internal spaces are folded determines the features we consider immutable laws of the universe—such as the charge on an electron or how gravity works. M-theory therefore allows for different universes with different laws—up to 10^{500} of them, in fact. Not only is there no formulation of M-theory, there is no consensus on what type of thing it is—a single theory, a network of connected theories, or something that changes according to circumstances? No-one is even very sure what the M stands for. What Anaxagoras called Mind (*nous*) and Einstein called a unified field theory may now be called M-theory, but we are little closer to knowing what that answer actually is—there is still plenty of physics left to be done.

The star cluster Messier 13, observed by Halley: "This is but a little Patch, but it shews itself to the naked Eye, when the Sky is serene and the Moon absent."

GLOSSARY

absolute zero The absolute minimum temperature possible (–459.4°F/–273°C), at which atomic motion stops completely.

aether The fifth element of ancient models of matter, after earth, air, fire, and water.

alchemy Trying to turn base metals into gold, or to find the elixir of life.

antimatter A form of matter in which each particle (such as an electron) has an opposite charge to its counterpart.

astrolabe An early astronomical tool that represents the position of the planets and stars, with the Earth at the center.

Big Bang Theory An explanation of the history of the universe that states that it began with a massive chemical reaction and we are still observing the results.

black hole An area of space-time with a gravitational force so high that no radiation (including light) can escape from it.

dark matter A term describing the 96 percent of the universe's matter that cannot be observed as light.

dynamics The study of movement and force.

electromagnetic radiation Transverse waves of energy moving at the speed of light.

elemental theory The theory that everything in the known world is made up of a number of pure substances, or elements.

empiricism Testing theories through experimentation and observation.

fluid mechanics The study of the movement of gases and liquids.

gluon Theoretical particle with no mass, believed to bind quarks together to form hadrons.

hadron A class of subatomic particle composed of quarks and taking part in strong interaction.

Higgs boson particle A hypothetical particle that has large mass and zero spin.

inertia The resistance to movement that non-moving objects exhibit, and a central core of Newton's study of dynamics.

isotope One of two or more forms of the same element that contain the same number of protons but different numbers of neutrons in their nuclei.

lepton Any of the group of subatomic particles (including the electron) that is not affected by the strong force.

kinetic energy The energy of a moving body.

momentum The tendency of a moving object to continue moving.

neutrino A subatomic particle with no electric charge, traveling through normal matter with little interaction.

neutron One of the basic subatomic particles having the same mass as a proton but no electric charge.

nuclear fission The splitting of an atomic nucleus into two parts.

Ohm's Law A law stating that electrical current is proportional to voltage and inversely proportional to resistance.

parallax The difference in the apparent position of an object when viewed from two different lines of sight.

photon A constantly moving bundle of electromagnetic energy, usually associated with light.

Planck's constant A figure that gives the unchanging ratio between the energy of a quantum of radiation and its frequency.

pulsar A magnetized, rotating stellar body whose rotation sends out pulses of radiation.

quasar A "quasistellar object" at the outer fringe of the known universe.

radioactive decay The emission of particles from the nucleus of an unstable isotope.

red shift The shift toward the red end of the spectrum in radiation from objects that are moving away from the observer.

spectroscopy A study of the characteristic patterns of lines on a spectrum.

string theory An mathematical explanation of how fundamental particles behave, suggesting that they are tiny fragments of "string" that vibrate in many dimensions.

strong force The interaction that binds protons and neutrons in the nucleus of an atom.

thermodynamics The relationship between heat and other forms of energy.

wave-particle duality (Of light and other forms of radiation) behaving sometimes like a wave and sometimes like a particle.

wave theory The theory that light and other forms of electromagnetic energy are made up of waves.

weak force The fundamental force that governs the interaction between hadrons and leptons.

For Further Reading

Close, Frank. *Particle Physics: A Very Short Introduction.* New York, NY: Oxford University Press, 2004

Holzner, Steven. *Physics for Dummies.* Hoboken, NJ: Wiley Publishing Inc, 2011

Pickover, Clifford A. *The Physics Book: From the Big Bang to Quantum Resurrection, 250 Milestones in the History of Physics.* New York, NY: Sterling, 2009.

Web Sites

Due to the changing nature of Internet links, Rosen Publishing has developed an online list of Web sites related to the subject of this book. This site is updated regularly. Please use this link to access the list:

http://www.rosenlinks.com/hos/phys

Index

Abhaya, King 64
absolute zero 92-3
Académie des Sciences 13
Academy of Experiment 11
aether 21, 22, 43-4, 47, 52-6
Age of Reason, The 25
al-Biruni, Abu Rayhan 9, 150, 152
al-Farisi, Kamal al-Din 38
al-Ghazali 23
al-Haytham, Ibn al-Hassan Ibn 8-9, 37-8, 39, 57, 150
al-Khwarizimi, Muhammed ibn Musa 149-50
al-Marwazi, Habash al-Hasib 150
Al-Rahwi 9
al-Shirazi, Qutb al-Din 38
al-Sufi, Abd al-Rahman 150
Albert of Saxony 68-9
alchemy 25, 26
Alhazen 8-9, 37-8, 39
Amontons, Guillaume 93
Ampère, André-Marie 91, 101
Analytical Mechanics (Lagrange) 80
Anaxagoras 16-18, 19, 60, 77, 189
Ancient Greece:
astronomy in 144-7; ideas of atoms 18-22; ideas of light 36-7; ideas of magnetism 100; ideas of matter 16-22; mechanics in 65-7; physics in 7-8
Anderson, Carl 135
Aniximander 7
antimatter 134-5
Aquinas, Thomas 68
Arab world:
astronomy in 148-50, 152; ideas of light 37-8; ideas of matter 23-4; ideas of mechanics 74; physics in 8-9
Archimedes 66
Aristarchus 144-5
Aristotle 8, 189:
ideas of matter 20, 21-2; ideas of mechanics 65-6, 67; ideas of time 184; medieval Europe 10, 24, 40; scientific method 13; speed of light 57
Arkwright, Thomas 79-80
armillary sphere 151
Arrhenius, Svante August 99, 102
Aryabhatiya 148
Aston, Francis 175-6
astrolobe 151
astronomy *see also* space-time continuum:
ancient Greece 144-7; Arab world 148-50, 152; brightness of stars 174-5; China 143; and comets 166-9; Copernican model 153-4, 160; energy in stars 175-6; Galileo 161-4, 165-6; India 147, 148; instruments for 151; Isaac Newton 164, 165, 166; Johannes Kepler 157-9; measurements in 166-7, 172-4; Mesopotamia 143-4, 151; Milky Way 174; prehistoric period 142; Ptolemaic model 147, 153, 154, 156; pulsars 180, 181; quasars 180; radio astronomy 176-80; spectroscopy 170-2; supernova viewed 152-3; telescopes 159-60, 161-2, 165-6, 173; Tycho Brahe 155-6, 157
atoms
17th-century ideas of 26-9; Ancient Greek ideas of 18-22; antimatter 134-5; Arab world ideas of 23-4; Bohr's ideas of structure 114-16; compounds 31-2; Copenhagen Interpretation 124-7; corpuscularianism 25; decay 128, 129-30; description of 19; dispute over existence of 33; electrons 112, 114-16, 118, 119, 120-3; elements 29-32; force on 128-9; Higgs boson particles 137, 139, 200-2; Indian ideas of 22-3; neutrinos 127, 135-7; neutrons 127-9; nuclear fission 131-2; plum pudding, model of 112-14; quantum mechanics 33, 116-29, 133-7, 139; thermodynamics 32-3
Aubrey, John 44, 48
Augustine, St. 186
Averroes 24, 68
Avicenna 9, 57, 67-8
Avogadro, Amedeo 32
Bacon, Francis 10, 57
Bacon, Roger 9-10, 38, 40-1, 57
Baliani, Batista 77
Becher, Johann 85
Becquerel, Alexandre 53, 106-7, 108, 129
Beeckman, Isaac 71
Bernoulli, Daniel 78-9, 88
Berti, Gasparo 77
Bessel, Friedrich 172
Bevis, John 153
Bhaskara 85
Big Bang theory 192-3, 195
black-body radiation 94-5, 177-8
black holes 187
Bohr, Niels 114-16, 117, 125, 138
Boltzmann, Ludwig 33, 114
Bonaparte, Napoleon 13
Book of Bodies and Distances (al-Marwazi) 150
Book of the Devil Valley Master 100
Bothe, Walter 127
Boyle, Robert 11, 12, 25, 27, 29, 49, 93
Bradley, James 59
Brady, Nicholas 29
Brahe, Tycho 151, 155-6, 157
Brahmagupta 148
Brentano, Franz 33
Broglie, Louis-Victor de 118, 119
Brown, Robert 32
Browne, Thomas 97
Bruno, Giordano 160
Buridan, Jean 68, 69, 153
Burnell, Jocelyn Bell 181
camera obscura 39
Cannon, Annie Jump 170, 171-2
Carnot, Nicolas Sadi 90-3
Cartesian geometry 45
Cassini, Giovanni 57-8, 166
Cavendish, Henry 98
celestial mechanics 75-6, 81
Cepheids 173-4
Cesi, Federico 11, 12
Chadwick, James 127-8, 130, 135
Châtelet, Émilie du 85, 87
China:
astronomy in 143; ideas of magnetism 100; mechanics in 65
Clausius, Rudolf 90, 91
comets 166-9
compounds 31-2
Copenhagen Interpretation 124-7
Copernicus, Nicolaus 153-4
Coriolis, Gustave-Gaspard de 96
corpuscularianism 25, 47-8
Cowan, Clyde 135-6
Curie, Marie 107-8, 109
Curie, Pierre 108, 109
Dalton, John 31-2, 93, 112
dark energy 199-200
dark matter 198-9
Davy, Humphry 88, 89, 105
De aspectibus (al-Haytham) 37
De magnete (Gilbert) 100, 145
De Nova Stella (Brahe) 156
deductive reasoning 8
Dee, John 160
Democritus 9, 18, 19, 21, 189
Desaulx, J. 32
Descartes, René:
Cartesian geometry 45; describes universe 190; ideas of matter 22, 25, 27, 27-8; ideas of mechanics 71, 72; ideas of optics 42, 43-4; speed of light 57, 60
DeWitt, Bryce 195
Dialogue of the Two Chief World Systems (Galileo) 163-4
Dicke, Robert 193
Digges, Leonard 43, 159
Digges, Thomas 159-60
Dirac, Paul 122-3, 134-5
Discourses and Mathematical Demonstrations Concerning Two New Sciences (Galileo) 164
Dokkum, Pieter van 194
Draper, Henry 170
Draper Catalogue of Stellar Spectra 170, 171-2
Dresden Codex 152
Du Shi 65
dynamics 67-72, 74
Eck, Johannes 11
Eddington, Arthur 60-1, 175, 176
Edison, Thomas 176
Einstein, Albert 121:
aether 22; Big Bang theory 192; Copenhagen Interpretation 125-7; direction of light 60; electromagnetism 103; EPR paradox 125-7; expanding universe 190-1; Manhattan Project 132; movement of molecules 32-3; photoelectric effect 52, 53; scientific method 13; speed of light 60; theory of relativity 186, 188; unified theory 202; wave-particle duality 119-20
electricity 96-101
electromagnetism 51-2, 61, 101-4
electrons 112, 114-16, 118, 119, 120-3
elemental theory 20-1, 85
elements 29-32
Empedocles 20, 21, 36, 47, 57, 85
empiricism 26
energy:
conservation of 84; conversion of 84-5, 89; dark energy 199-200; electricity 96-101; electromagnetism 51-2, 61, 101-4; heat 85-6; kinetic energy 96; magnetism 100-1; perpetual motion 85; potential energy 96; radiation 106-9; thermodynamics 88-93; X-rays 104-5, 106-7
Epicurus 18-19
EPR paradox 125-7
Euclid 36
Euler, Leonard 80
Everett III, Hugh 125, 195
Exner, Felix Maria 32
Fabricius, Johann 162

Faraday, Michael 51, 90, 101, 102, 104, 105
Fay, Charles du 99
Fermat, Pierre de 60
Fermi, Enrico 130, 131, 132, 135
Feynmann, Richard 84, 133, 137, 138
fire 85-6
Fizeau, Hippolyte 59
Flammarion, Camille 169
Flamsteed, John 164-5
Fleming, Williamina 170-1
fluid mechanics 76-9, 80
Foucault, Léon 59-60
Foucault's pendulum 13
Franklin, Benjamin 97-8, 99
Fresnel, Augustin-Jean 50
Friedmann, Alexander 190-1
Fujiwara, Sadiae 153
Galilei, Galileo 6, 73:
conflict with Church 162-4; ideas of mechanics 71-2, 77; measurement of time 184; member of Lyncean Academy 11; rolling-ball experiment 71-2; scientific method 10-11; speed of light 57; telescope 41, 43, 73, 161-2
Galileo (spacecraft) 160
Galvani, Luigi 99
Gamow, George 192, 193
Gan De 161
Gassendi, Pierre 25, 47, 74, 156
Gay-Lussac, Joseph Louis 91, 93
Geber 9
Geiger, Hans 112
Gell-Mann, Murray 134
Gilbert, William 100-1, 145
gluons 133
Gouy, Louis Georges 32
Gravesande, Willem 85
gravity 75, 159
Gray, Stephen 98-9
Grosseteste, Richard
Guericke, Otto von 27, 31, 97
Gurzadyan, Vahe 195
Halley, Edmund 164, 165, 166-8
Halley's comet 168-9
Hamilton, Sir William Rowan 80, 81
Hargreaves, James 79
Harroit, Thomas 162
Hartsoeker, Nicolaas 28-9
Hawking, Stephen 33, 202, 203
heat:
absolute zero 92-3; caloric model 88-9, 91; ideas of fire 85-6; light 93-5; mechanical model 88, 89-90; thermodynamics 88-93
Heaviside, Oliver 178
Heisenberg, Werner 122-3, 124
Henderson, Thomas 172
Henry, Joseph 101-2, 103
Hero of Alexandria 36-7, 77
Hertz, Heinrich 33, 104, 107
Hertzsprung, Ejnar 173, 174-5
Hertzsprung-Russell diagram 174, 175
Hewish, Antony 179
Hey, James 179
Higgs, Peter 201
Higgs boson particle 137, 139, 200-2
Hipparchus 67, 145-6
Hippocrates 20
Hittorf, Johann Wilhelm 104
homoiomeries 19
Hooke, Robert 11, 12, 49, 164, 184; astronomy 172; ideas of light 45; and Isaac Newton 46-7, 49; Micrographia 11, 12, 46-7; microscopy 46-7
Horrocks, Jeremiah 167
Hoyle, Fred 193
Hubble, Edwin 174, 191, 193
Hubble Space Telescope 173
Huygens, Christiaan 43:
ideas of light 48, 50, 57, 58; pendulum clock 184; telescope 165-6
Hydrodynamica (Bernoulli) 78, 88
impetus 67-70
India:
astronomy in 147, 148; ideas of atoms 22-3; ideas of light 36
inductive reasoning 8
inertia 72, 74
Jansky, Karl 178-9
Joliot-Curie, Frédéric 126, 127, 131, 132
Joliot-Curie, Irène 126, 127, 131
Joule, James Prescott 89-90, 93
Kanada (Kashyapa) *22, 23*
Karlsruhe Tritium Neutrino Experiment (KATRIN) 137
Kelvin, Lord 54, 90, 93
Kennelly, Edwin 178
Kepler, Johannes 41-2, 68, 73, 75, 157-9
kinetic energy 96
Kleist, Georg von 97
Lagrange, Joseph-Louis 31, 80
Lambert, Heinrich 93
Laplace, Pierre Simon 76, 93, 187
Lattes, César 128
Lavoisier, Antoine 16, 29-31, 86, 88-9, 93
Leavitt, Henrietta Swan 171, 173
Leeuwenhoek, Antonie van 43
Leibniz, Gottfried 76, 85
Leigh, David 92
Lemaître, Georges 192-3
lenses 43
Leucippus 18-19, 189
Leyden jars 97
light *see also* optics:
16th- and 17th-century ideas of 41-4; aether 43-4, 47, 52-6; Arab world ideas of 37-8; electromagnetic radiation 51-2, 61; heat 93-5; Indian ideas of 36; Isaac Newton's ideas on 47-8; Medieval European ideas of 38-41; photoelectric effect 52, 53; photons 52, 53; speed of 57-60, 188-9; straight line 60-1; wave-particle duality 117-27; wave theory development 49-52
Lodge, Sir Oliver 176-7
Lorentz, Hendrik 22
Lyncean Academy 11-12, 94
M-theory 203
Mach, Ernst 81
Magiae naturalis (della Porta) 94
Manhattan Project 132
Marrison, Warren 184
mass, conservation of 16, 30
matter:
Anaxagoras 16-18, 19; Ancient Greek ideas of 16-22; Arab world ideas of 23-4; atomism 18-24; elemental theory 20-1; Indian ideas of 22-3; medieval European ideas of 21, 24-5
Maury, Antonia 171
Maxwell, James Clerk 22, 51-2, 90, 92, 102-3, 106, 166
Mayans 152
mechanics:
16th and 17th century ideas of 70-2; 18th- and 19th-century ideas of 80-1; Ancient Greece 65-7, 68; Arab world ideas of 74; celestial mechanics 75-6, 81; China 65; dynamics 67-72, 74; fluid mechanics 76-9, 80; Galileo 71-2; impetus 67-70; Industrial Revolution 79-80; inertia 72, 74; Isaac Newton 74-5; Mesopotamia 64-5; momentum 72, 74; Oxford Calculators 70; Sri Lanka 64; static mechanics 68; tunnel experiment 69-70; velocity 70
medieval Europe:
Aristotle's influence on 10, 24, 40; ideas of light 38-41; ideas of matter 21, 24-5; ideas of mechanics 70; physics in 9-10
Mendeleev, Dmitri 30
Mersenne, Marin 57
mesons 128
Mesopotamia 64-5, 143-4, 151
Michell, John 187
Michelson, Albert 54-6
Micrographia (Hooke) 12, 46-7, 49
Milky Way 174
Mill, John Stuart 42
momentum 72, 74
Morley, Edward 54-6
motion *see* mechanics
Muller, Johannes 153
multiverses 195
Musschenbroek, Pieter van 97
Nagaoka, Hantaro 114
negative refractive index 59
Nernst, Walther 92
neutrinos 127, 135-7
neutrons 127-9
New Experiments and Observations Touching Cold (Boyle) 93
New Organon of the Sciences, The (Bacon) 10
Newton, Isaac 76:
astronomy 164, 165, 166; corpuscularianism 25; deductive and inductive reasoning 8; describes universe 190; electricity 97; gravity 75, 159; ideas of atoms 28, 31; ideas of light 47-8, 58-9; ideas of mechanics 74-6; ideas of optics 44-7; ideas of time 185; laws of motion 74-5; Robert Hooke 46-7, 49
Nicholas of Autrecourt 24-5
Nordman, Charles 177
nuclear fission 131-2
Occhialini, Giuseppe 128
Ohm, Georg 99
Ohm's law 99-100
On a General Method of Dynamics (Hamilton) 80
On the Revolution of the Celestial Spheres (Copernicus) 153-4
optics *see also* light:
Isaac Newton's ideas on 44-8; lens development 43; René Descartes ideas on 42, 43-4; Robert Hooke's work on 46-7
Optics (Bacon) 40
Oresme, Nicole 186
Ørsted, Christian 101
Oxford Calculators 70
Parakrambahu, King 64
parallax 171, 172, 173
Parmenides 16, 21
partons 133-4
Pascal, Blaise 78
Pauli, Wolfgang 127, 135, 136
Peebles, Jim 193
Penrose, Sir Roger 195
Penzias, Arno 193
Perlmutter, Saul 200
perpetual motion 85
Perrin, Jean 33, 112
Philoponus, John 67
Philosophical Transactions 12
phlogiston 85-6
photoelectric effect 52, 53
physics:
Ancient Greece ideas of 7-8; Arab world ideas of 8-9; description of 6-7; growth of scientific societies 11-13; medieval Europe ideas of 9-10
Pickering, Edward 170-1
Pingzhou Table Talks (Zhu Yu) 100
Planck, Max 52, 53, 96:
black-body radiation 94-5; 177-8; quantum mechanics 116
Planck's constant 118-19

Plato 8:
ideas of atoms 18, 20, 21; ideas of light 36; ideas of time 184
Pliny the Elder 43
plum pudding atomic model 112-14
Podolsky, Boris 125
Poisson, Siméon-Denis 91
Pope, Alexander 44
Porta, Giambattista della 11, 94
positron 135
potential energy 96
Powell, Cecil 128
Prévost, Pierre 89
Principa (Newton) 164, 167
Prognostication Everlasting (Digges) 160
Pseudo-Geber 24
Ptolemy, Claudius 37, 60, 146, 147, 151
pulsars 180, 181
Pythagoras 7, 144
quadrant 151
quantum mechanics 33, 116-29, 133-7, 139
quarks 133
quasars 180, 194-5
radiation 106-9
radio astronomy 176-80
Ramsay, Sir William 130
Rankine, William 96
rationalism 26
Reber, Grote 179
red shift 191
refraction of light 60-1
Regiomantus 153
Reimann, Bernhard 188
Reines, Frederick 135-6
Richer, Jean 166
Ridwan, Ali ibn 150
Rohault, Jacques 28
rolling-ball experiment 71-2
Rømer, Ole 57-8
Röntgen, Wilhelm Conrad 104-5, 108
Rosen, Nathan 125
Royal Observatory 164
Royal Society 12
Rudolph II, Emperor 157, 158
Rumford, Benjamin Count Rumford 89
Russell, Henry 174-5
Rutherford, Ernest 108, 112, 114, 129-30
Ryle, Martin 179
Sahl, Ibn 60
Samkhya school 36
Sceptical Chymist, The (Boyle) 25
Scheiner, Julius 177
Schmidt, Maarten 180
Schwarzschild, Karl 187
scientific method 8, 9, 10-11, 13, 26, 39
scientific societies 11-13
Shakir, Ja'far Muhammad ibn Musa ibn 150
Shapley, Harlow 174
Shrödinger, Erwin 120-3, 124-5
Skellett, Albert 178
Slipher, Vesto 191
Snellius, Willebrord 60
Snell's Law 60
Socrates 17, 18
Soddy, Frederick 128, 129-30
space-time continuum *see also* astronomy:
Big Bang theory 192-3, 195; black holes 187; dark energy 199-200; dark matter 198-9; expanding universe 190-2; multiverses 195; number of stars 193-5; origins of the universe 189-90; red shift 191; speed of light 188-9; theory of relativity 186, 188
spectroscopy 170-2
Sri Lanka 64
Stahl, Georg Ernst 85-6
Starry Messenger (Galileo) 162, 163
stars:
brightness of 174-5; Cepheids 173-4; distances of 172-4; energy in 175-6; number of 193-5; and spectroscopy 170-2, 191
Stonehenge 142
string theory 202-3
subatomic particles *see* quantum mechanics
Szilárd, Leó 131-2
telescopes 41, 42-3, 73, 159-60, 161-2, 165-6, 173
Tesla, Nikola 177
Thales of Miletus 7, 8, 16, 100
thermodynamics 32-3, 88-93
Thomson, George 119
Thomson, Joseph John 33, 112-14
time:
measurement of 184-6; Zeno paradox 184, 185
Torricelli, Evangelista 12, 78
Twain, Mark 169
unified theory 202-3
Ussher, James 189
vacuum, discovery of 27
Vaisheshika school 36
Vasaba, King 64
Vedanga Jyotisa 148
velocity 70
Vishnu Purana 36
Viviani, Vincenzo 12
Volta, Alessandro 99
wave-particle duality 117-27
wave theory 49-52
Wheeler, John Archibald 187
Wilkinson, David 193
Wilsing, Johannes 177
Wilson, Robert 193
Wren, Sir Christopher 12, 164
Wright, Edward 145
X-rays 104-5, 106-7
Young, Thomas 51
Yukawa, Hideki 128-9
Zeno paradox 184, 185
Zhang Heng 65
Zhu Yu 100
Zij al-Sindh (al-Khwarizimi) 149
Zwicky, Fritz 198, 199

Picture Credits

Shutterstock: 6, 7, 14, 17 (x2 btm), 18, 21 (x2 top), 41 (l), 55 (top), 57, 59, 68, 84 (btm), 85, 116 (btm), 117 (top), 134, 142 (top), 149 (btm), 151 (all), 155, 171
Photos.com: 11, 17 (top), 20, 26
Corbis: 34, 56 (btm), 61, 82, 109, 110, 119, 127, 146 (btm), 192, 196
Bridgeman: 21 (btm), 30, 44, 62, 162 (btm), 168 (btm)
Science Photo Library: 29, 31, 37 (top), 53 (top), 54, 76 (top), 88, 92, 96 (btm), 101, 122 (btm), 125 (top), 126 (top), 126 (btm), 135, 145, 162 (mid), 173 (btm); 174 (top), 174 (btm), 179, 180, 181 (btm), 182, 187, 190, 191
British Museum: 43 (btm)
Mary Evans: 147
Topfoto: 66 (l), 79, 81, 149 (top), 161
Getty: 72
Clipart: 67, 90
The Nuremberg Chronicle: 8; Rita Greer: 12; Arnaud Clerget: 13 (top); *A Pictorial History* by Joseph G Gall (1996): 13 (btm); Rebecca Glover: 23, 47, 170 (top); Monfredo de Monte Imperiali, 14th C: 24; Johann Kerseboom: 27; Smithsonian Institution (The Dibner Library Portrait Collection): 33; Girolamo di Matteo de Tauris for Sixtus IV: 36; Nino Pisano: 37 (btm); Noé Lecocq 38 (top); bank note Iraq: 38; Gnangarra: 39 (top); Stefan Kühn: 39 (btm), 129; Bibliotheca Apostolica Vaticana: 40; René Descartes: 42 (r); Isaac Newton: 45 (btm), 47 (top), 74, 165; Hooke: 48; Wenceslas Hollar: 49; Christiaan Huygens: 50 (top); Caspar Netscher: 50 (btm); James Clerk Maxwell 52 (btm); Alain Le Rille: 55 (btm); Falcorian; 56 (top); Giulio Parigi: 58; Ibn Sahl's manuscript: 60; Antoine-Yves Goguet: 64; Tamar Hayardeni: 66 (r); Bill Stoneham: 70; Justus Sustermans: 73; Sir Godfrey Kneller: 76 (btm); Danielis Bernoulli: 78; Nicolas de Largillière: 87; *Harper's New Monthly Magazine*, no.231, Aug 1869: 89; V Bailly (1813): 91; Peter Gervai: 93; ESA/NASA: 94, 178 (top), 188; Scott Robinson: 96 (top); Otto von Guericke: 97 (top); L. Margat-L'Huillier, *Leçons de Physique* (1904): 97 (btm); *Natural Philosophy for Common and High Schools* (1881), p.159 by Le Roy C Cooley: 98; Ryan Somma: 100; National Archaeological Museum of Spain: 100 (btm); Nevit Dilmen: 103 (btm); *Experimental Researches in Electricity* (vol.2, plate 4): 103 (top); Arthur William Poyser (1892) *Magnetism and Electricity*: 104; Harriet Moore: 105; George Grantham Bain Collection (Library of Congress): 108, 175 (top); NASA/JPL: 115 (top), 166; Paul Ehrenfest: 117 (btm); inductiveload: 118; US Airforce: 120; Gerhard Hund: 122 (top); Berlin Robertson: 125 (btm); Smithsonian Institute: 130; Gary Sheehan (Atomic Energy Commission: 131 (btm); Julian Herzog; 139; ESO/Stéphane Guisard; Simon Wakefield: 142 (btm); Brian J Ford, *Images of Science* (1993): 144 (btm); J van Loon (*ca.*1611–1686): 146 (top); NASA: 150, 202; Dresden Codex: 152; Andreas Cellarius: 154; *Die Gartenlaube*: 158; Whipple Museum of the History of Science: 159 (r); Thomas Murray: 164; Antonio Cerezo, Pablo Alexandre, Jesús Merchán, David Marsán: 167; The Yerkes Observatory: 168 (mid); *Popular Science Monthly*, Vol. 11: 169; Smithsonian Institute Archives: 170 (btm); Michael Perryman: 172; NASA/HST: 173 (top); Hannes Grobe: 175 (btm); Napoleon Sarony: 177; Yebes: 178 (btm); Astronomical Institute, Academy of Sciences of the Czech Republic: 181 (top); Museo Barracco: 184; Sandro Botticelli: 186; NASA/WMAP Science Team: 194 (top), 200; NASA/Swift/S Immler: 194 (btm); NASA/ESA, MJ Lee and H Ford (Johns Hopkins University): 198; Lucas Taylor: 201; ESA/Hubble and NASA: 203
Wikiuser: Didier B 71; Tamorlan 77 (btm); rama 86 (top); Fastfission: 114; orangedog: 116 (top); ShakataGaNai: 136
Unknown author: 25, 41, 42 (l), 51 (l), 52 (top), 99, 115, 153, 156 (r)